# Learn Ansible

Automate your cloud infrastructure, security configuration,
and application deployment with Ansible

**Russ McKendrick**

# Learn Ansible

**Group Product Manager**: Preet Ahuja
**Publishing Product Manager**: Suwarna Rajput
**Book Project Manager**: Ashwini Gowda
**Senior Editor**: Runcil Rebello
**Technical Editor**: Arjun Varma
**Copy Editor**: Safis Editing
**Proofreader**: Runcil Rebello
**Indexer**: Subalakshmi Govindhan
**Production Designer**: Ponraj Dhandapani
**DevRel Marketing Coordinator**: Rohan Dobhal

First published: July 2018

Second edition: June 2024

Production reference: 1080524

Published by Packt Publishing Ltd.

Grosvenor House
11 St Paul's Square
Birmingham
B3 1RB, UK

ISBN 978-1-83508-891-3

www.packtpub.com

# Contributors

## About the author

**Russ McKendrick** is an experienced DevOps practitioner and system administrator with a passion for automation and containers. He has been working in IT and related industries for the better part of 30 years. During his career, he has had responsibilities in many different sectors, including first-line, second-line, and senior support in client-facing and internal teams for small and large organizations.

He works almost exclusively with Linux, using open source systems and tools across dedicated hardware and virtual machines hosted in public and private clouds at Node4, where he is the practice manager (**site reliability engineering (SRE)** and DevOps). He also buys way too many records!

*I would like to thank my family and friends for their support and for being so understanding about all of the time I have spent in front of the computer writing. I would also like to thank my colleagues at Node4 and our customers for their kind words of support and encouragement throughout the writing process.*

# About the reviewer

**Luca Berton** is a seasoned Ansible Automation expert and a pivotal member of the Red Hat Ansible engineer team, where he honed his skills for three years. He is the celebrated author of several Ansible best-selling books. His infrastructure hardening and automation expertise is backed by over 15 years of experience integrating with Kubernetes and Terraform.

Luca created the Ansible Pilot website and YouTube channel. His passion for open source solutions is evident through his active participation in community events, where he shares extensive knowledge to help others excel in automation.

Luca continues to contribute to the technological community, driving innovations and simplifying automation challenges for professionals around the globe.

# Table of Contents

# Part 2: Deploying Applications

## 4

## 5

# 6

## Targeting Multiple Distributions    135

# 7

## Ansible Windows Modules    151

# Part 3: Network and Cloud Automation

# 8

## Ansible Network Modules    171

# Part 4: Ansible Workflows

# Preface

Ansible, an open source orchestration tool, has experienced significant growth and is now a comprehensive orchestration and configuration management solution under Red Hat's ownership. This book will guide you through writing playbooks using core, vendor-supplied, and community Ansible modules to deploy various systems, from simple LAMP stacks to highly available public cloud infrastructures.

By the end of this book, you will have acquired the following skills:

- A solid foundational knowledge of Ansible and its various supporting tools

- The ability to write your own custom playbooks to configure both Linux and Windows servers

- The ability to define highly available cloud infrastructures using code, enabling the easy distribution of your infrastructure configuration alongside your code base

- An understanding of how to use Ansible Galaxy, use community-contributed roles, and create and contribute your own roles

- The ability to run your Ansible playbooks using GitHub Actions and Azure DevOps

- The ability to deploy and configure Ansible AWX, a web-based interface for Ansible

- Various skills gained from exploring several use cases demonstrating how to integrate Ansible into your daily tasks and projects

You will have a solid understanding of how to incorporate Ansible into your everyday responsibilities as a system administrator, developer, or DevOps practitioner.

## Who this book is for

This book is written for people in the following roles who want to streamline their workflows by leveraging Ansible's capabilities:

- **System administrators**: This book will help you automate repetitive tasks and ensure consistent configurations across your systems if you manage and maintain servers, networks, and other infrastructure components.

- **Developers**: As a developer, you can benefit from this book by learning how to use Ansible to provision and manage development environments, deploy applications, and integrate infrastructure-as-code practices into your development workflow.

- **DevOps practitioners**: If you are a DevOps practitioner responsible for bridging the gap between development and operations, this book will provide the tools and knowledge to create efficient, repeatable, and scalable deployment processes using Ansible.

No prior experience with Ansible is necessary to get started with this book.

## What this book covers

*Chapter 1, Installing and Running Ansible*, discusses the problems Ansible was developed to solve. After covering its background, we will work through installing Ansible on macOS and Linux. We will also discuss why there is no native Windows installer and cover installing Ansible on the Windows Subsystem for Linux.

*Chapter 2, Exploring Ansible Galaxy*, discusses Ansible Galaxy, an online repository of community and vendor-contributed roles. In this chapter, we will discover some of the best roles available, how to use them, and how to create your role and have it hosted on Ansible Galaxy.

*Chapter 3, The Ansible Commands*, explains how we examine Ansible commands before writing and executing more advanced playbooks. Here, we will cover using the tools that make up Ansible.

*Chapter 4, Deploying a LAMP Stack*, discusses deploying a complete LAMP stack using the various core modules that ship with Ansible. We will target the Ubuntu machine that is running locally.

*Chapter 5, Deploying WordPress*, expands on the LAMP stack playbook, which we deployed in the previous chapter as our base. We will use Ansible to download, install, and configure WordPress – a popular CMS.

*Chapter 6, Targeting Multiple Distributions*, explains how we will adapt the playbook from the previous chapter so it can run against both Debian, which we have been targeting so far, and Red Hat-based Linux distributions.

*Chapter 7, Ansible Windows Modules*, explores the ever-growing collection of Ansible modules that support and interact with Windows-based servers.

*Chapter 8, Ansible Network Modules*, discusses network modules available from various vendors through Ansible Galaxy. Due to their requirements, we will only discuss the functionality of these modules.

*Chapter 9, Moving to the Cloud*, discusses how we can move from using local virtual machines to using Ansible to deploy network and compute resources in Microsoft Azure. Then, we will use the playbook from the previous chapters to install and configure a LAMP stack and WordPress.

*Chapter 10, Building Out a Cloud Network*, since we will have just launched a virtual machine in Microsoft Azure, moves on to Amazon Web Services; however, before launching any compute instances, we must create a network in which they can be hosted.

*Chapter 11, Highly Available Cloud Deployments*, continues our Amazon Web Services deployment. We will start deploying compute and storage services into the network we created in the previous chapter, and by the end of the chapter, we will have a highly available WordPress installation.

*Chapter 12, Building Out a VMware Deployment*, discusses the modules that allow you to interact with the various components of a typical VMware installation.

*Chapter 13, Scanning Your Ansible Playbooks*, provides practical examples of running two third-party tools, Checkov and KICS. These tools are designed to scan your Ansible playbook code for common mistakes and potential security problems.

*Chapter 14, Hardening Your Servers Using Ansible*, explains how to install and execute OpenSCAP. We will also automatically generate remediation Ansible playbooks and Bash scripts to resolve any problems found during the scan. We will also look at running WPScan and OWASP ZAP scans against the resources deployed using the playbooks from previous chapters.

*Chapter 15, Using Ansible with GitHub Actions and Azure DevOps*, will examine running our Ansible playbook books from these two CI/CD platforms. As neither has native Ansible support, we will discuss how to install and run Ansible to get the most out of the platforms.

*Chapter 16, Introducing Ansible AWX and Red Hat Ansible Automation Platform*, examines two web-based interfaces: we will discuss the commercial Red Hat Ansible Automation Platform and then take a deep dive into deploying and configuring the open source Ansible AWX.

*Chapter 17, Next Steps with Ansible*, discusses how Ansible can be integrated into our day-to-day workflows, from interacting with collaboration services to troubleshooting your playbooks with the built-in debugger. We will also look at real-world examples of how I have used Ansible across organizations I have worked with.

## To get the most out of this book

To get the most out of this book, I assume that you have the following:

- Some experience of using the command line on both Linux-based machines and macOS
- A basic understanding of how to install and configure services on a Linux server
- A working knowledge of services and languages such as Git, YAML, and virtualization

| Software/hardware covered in the book | Operating system requirements |
|---|---|
| Ansible | macOS, Linux, or Windows via Subsystem for Linux |
| Canonical Multipass | macOS, Linux, or Windows |
| Various CLIs for public cloud providers | macOS, Linux, or Windows via Subsystem for Linux |

If you are using the digital version of this book, we advise you to type the code yourself or access the code from the book's GitHub repository (a link is available in the next section). Doing so will help you avoid any potential errors related to the copying and pasting of code.

## Download the example code files

You can download the example code files for this book from GitHub at http://github.com/PacktPublishing/Learn-Ansible-Second-Edition. If there's an update to the code, it will be updated in the GitHub repository.

We also have other code bundles from our rich catalog of books and videos available at https://github.com/PacktPublishing/. Check them out!

## Conventions used

There are a number of text conventions used throughout this book.

Code in text: Indicates code words in text, database table names, folder names, filenames, file extensions, pathnames, dummy URLs, user input, and Twitter handles. Here is an example: "As you can see, it is calling a variable called {{ apache_packages }}, which is defined in roles/apache/defaults/main.yml as follows."

A block of code is set as follows:

```
- name: "Install apache packages"
  ansible.builtin.apt:
    state: "present"
    pkg: "{{ apache_packages }}"
```

When we wish to draw your attention to a particular part of a code block, the relevant lines or items are set in bold:

```
- name: "Install apache packages"
  ansible.builtin.apt:
    state: "present"
    pkg: "{{ apache_packages }}"
```

Any command-line input or output is written as follows:

```
$ ansible-playbook -i hosts site.yml
```

**Bold**: Indicates a new term, an important word, or words that you see onscreen. For instance, words in menus or dialog boxes appear in **bold**. Here is an example: "You can leave the remaining options at their defaults and then click on the **Create repository** button at the end of the form."

> **Tips or important notes**
> Appear like this.

## Get in touch

Feedback from our readers is always welcome.

**General feedback**: If you have questions about any aspect of this book, email us at customercare@packtpub.com and mention the book title in the subject of your message.

**Errata**: Although we have taken every care to ensure the accuracy of our content, mistakes do happen. If you have found a mistake in this book, we would be grateful if you would report this to us. Please visit www.packtpub.com/support/errata and fill in the form.

**Piracy**: If you come across any illegal copies of our works in any form on the internet, we would be grateful if you would provide us with the location address or website name. Please contact us at copyright@packt.com with a link to the material.

**If you are interested in becoming an author**: If there is a topic that you have expertise in and you are interested in either writing or contributing to a book, please visit authors.packtpub.com.

## Share Your Thoughts

Once you've read *Learn Ansible*, we'd love to hear your thoughts! Scan the QR code below to go straight to the Amazon review page for this book and share your feedback.

https://packt.link/r/1835088910

Your review is important to us and the tech community and will help us make sure we're delivering excellent quality content.

# Download a free PDF copy of this book

Thanks for purchasing this book!

Do you like to read on the go but are unable to carry your print books everywhere?

Is your eBook purchase not compatible with the device of your choice?

Don't worry, now with every Packt book you get a DRM-free PDF version of that book at no cost.

Read anywhere, any place, on any device. Search, copy, and paste code from your favorite technical books directly into your application.

The perks don't stop there, you can get exclusive access to discounts, newsletters, and great free content in your inbox daily

Follow these simple steps to get the benefits:

1.  Scan the QR code or visit the link below

https://packt.link/free-ebook/9781835088913

2.  Submit your proof of purchase
3.  That's it! We'll send your free PDF and other benefits to your email directly

# Part 1: Introducing, Installing, and Running Ansible

In this part, we will dive into the world of Ansible and explore the fundamental concepts. You will learn how to install Ansible on various operating systems and familiarize yourself with the basic commands and structure of Ansible playbooks. By the end of this part, you will have a solid foundation to build upon as we delve deeper into automating tasks with Ansible.

This part has the following chapters:

- *Chapter 1, Installing and Running Ansible*
- *Chapter 2, Exploring Ansible Galaxy*
- *Chapter 3, The Ansible Commands*

# 1

# Installing and Running Ansible

Welcome to this, our first chapter in the second edition of *Learn Ansible*. In this chapter, we will look at a few topics to introduce you to **Ansible**; these topics will familiarize you with the basics of what Ansible is and give you a few different use cases.

By the end of the chapter, you will have gotten hands-on with Ansible and covered the following:

- Who is behind Ansible?
- The differences between Ansible and other tools
- The problem Ansible solves
- How to install Ansible on macOS and Linux
- Running Ansible on Windows 11 using the Windows Subsystem for Linux
- Launching a test virtual machine
- An introduction to playbooks

Before we start talking about Ansible, let's quickly discuss my background, how I came to be writing a book about Ansible, and what you will need to install and run Ansible on your system.

## Technical requirements

Later in this chapter, we will install Ansible, so you will need a machine capable of running it. I will go into more detail about these requirements in the second half of the chapter. We will also use **Multipass** to launch a virtual machine locally. A section walks through installing Multipass and downloading an Ubuntu image to use the virtual machine's base, which is a download of a few hundred MBs. You can find all of the code used in this chapter at `https://github.com/PacktPublishing/Learn-Ansible-Second-Edition/tree/main/Chapter01`.

# My story: part one

I have been working with servers, primarily ones that serve web pages, since the late 90s, and the landscape is unrecognizable. Here is a quick overview of my first few years running servers to give you an idea of how I used to operate my early servers.

Like most people at the time, I started with a shared hosting account where I had very little control over anything on the server side when the website I was running outgrew shared hosting due to the forum, which made up part of the site's popularity. I moved to a dedicated server, where I thought I could flex my future system administrator muscles, but I was wrong.

The server I got was a Cobalt RaQ 3; this was a 1U server appliance that was ahead of its time. However, I did not have root-level access to the machine, and I had to use the web-based control panel for everything I needed to do. Eventually, I got a level of access where I could access the server using Telnet; I now know this isn't good, but it was the early days, and SSH was considered cutting-edge. I started to teach myself how to be a system administrator by making changes in the web control panel and looking at the changes to the configuration files on the server.

After a while, I changed servers and, this time, opted to forego any web-based control panel and use what I had learned with the Cobalt RaQ to configure my first proper **Linux**, **Apache**, **MySQL**, **PHP** (or **LAMP** for short) server by using the pages of notes I had made. I had created runbooks of one-liners to install and configure the software I needed and numerous scribbles to help me investigate problems and keep the lights on.

After I got my second server for another project, I realized that was probably a good time to type out my notes so that I could copy and paste them when I needed to deploy a server; the timing of this couldn't have been better as it was a few months after making these notes that my first server failed—my host apologized and replaced it with a higher-specification but completely fresh machine with an updated operating system.

So, I grabbed my Microsoft Word file containing my notes and copied and pasted each instruction, making tweaks based on what I needed to install on the upgraded operating system. Several hours later, I had my server up and running and my data restored.

One of the critical lessons I learned, other than that there is no such thing as too many backups, was not to use Microsoft Word to store these types of notes; the Linux command line doesn't care if your notes are all nicely formatted with headings and courier font for the bits you need to paste in. It does care about proper syntax, and Word had very kindly autocorrected and formatted all of my notes for print, meaning that not only did I have the pressure of having to deploy a new server and restore the backups I had thankfully been taking each day but also to try and debug my notes as I was doing so.

Because of this, I made a copy of the history file on the server and transcribed my notes in plaintext. These notes provided the base for the next few years as I started to script parts of them, mainly the bits that didn't require any user input.

These scraps of commands, one-liners, and scripts were all adapted through Red Hat Linux 6; note the lack of the word Enterprise appended to the operating system's name there, all the way through to CentOS 3 and 4.

Things got complicated when I changed roles; I stopped consuming services from a web hosting company and started working for one. Suddenly, I was building servers for customers who may have different requirements than my projects—each server was different.

From here, I started working with Kickstart scripts, PXE boot servers, gold masters on imaging servers, virtual machines, and bash scripts that started prompting information on the system being built. I had also moved from only needing to worry about maintaining my servers to having to log in to hundreds of different physical and virtual servers, from ones that belonged to the company I was working for to customer machines.

Over the next few years, my single text file quickly morphed into a complex collection of notes, scripts, precompiled binaries, and spreadsheets of information that only made sense to me; if I am being honest, I ended up making myself quite a significant single point of failure.

While I had moved to automate quite a few parts of my day-to-day work using bash scripts and stringing commands together, I found that my days were still very much filled with running all these tasks manually and working a service desk dealing with customer-reported problems and queries.

My story is typical of many people, while the operating systems used will probably be considered ancient. The entry point of using a GUI and moving to the command line while keeping a scratch pad of common commands is quite a common scenario I have heard when working with other system administrators and even modern-day DevOps practitioners.

So now that you know a little about my background, let's talk about Ansible.

# Ansible's story

Let's take a quick look at who developed Ansible and what it actually is.

## What is Ansible?

Before we discuss how Ansible started, we should quickly discuss the origin of the name. The term "Ansible" was penned by science fiction novelist Ursula K. Le Guin; it was first used in her novel *Rocannon's World*, which was first published in 1966. In the story's context, an Ansible is a fictional device that sends and receives messages faster than light.

> **Note**
> In 1974, Ursula K. Le Guin's novel *The Dispossessed: An Ambiguous Utopia* was published. This book features the development of the Ansible technology by exploring the (fictional) details of the mathematical theory that would make such a device possible.

The term has since been used by several other notable authors within the genre to describe communication devices that are capable of relaying messages over interstellar distances, which, as you will discover throughout the course of the book, is quite an apt description of the software itself.

## Ansible, the software

Ansible was initially developed by Michael DeHaan, who was also the author of **Cobbler**, which was developed while DeHaan worked for Red Hat.

> **Note**
>
> Cobbler is a Linux installation server that allows you to deploy servers within your network quickly; it can help with DNS, DHCP, package updates and distribution, virtual machine deployment, power management of physical hosts, and also the handoff of a newly deployed server, be it physical or virtual, to a configuration management system.

DeHaan left Red Hat and worked for companies such as **Puppet**, which was a good fit since many users of Cobbler used it to hand off to a Puppet server to manage the servers once they had been provisioned, myself included.

A few years after leaving Puppet, DeHaan made the first public commit to the Ansible project on February 23, 2012. The original README file gave quite a simple description that laid the foundation for what Ansible would eventually become:

> *Ansible is an extra-simple Python API for doing 'remote things' over SSH. As Func, which I co-wrote, aspired to avoid using SSH and have its own daemon infrastructure, Ansible aspires to be quite different and more minimal, but still able to grow more modularly over time.*

Since that first commit, and at the time of writing, there have been over 53,000 commits by 5,000 contributors, and the project has over 58,000 stars on GitHub.

In 2013, the project had grown. Ansible, Inc. was founded to offer commercial support to users who had relied on the project to manage their infrastructure and server configuration, whether physical, virtual, or hosted on public clouds.

Out of the formation of Ansible, Inc., which received $6 million in series A funding, came the commercial Ansible Tower, which acted as a web-based frontend where end users could consume role-based access to Ansible services.

Then, in October 2015, Red Hat announced they would acquire Ansible for $150 million.

In the announcement, Joe Fitzgerald, who was vice president, Management, at Red Hat at the time of the acquisition, stated, *"Ansible is a clear leader in IT automation and DevOps, and helps Red Hat take a significant step forward in our goal of creating frictionless IT."*

During this book, you will find that the statements in the original README file and Red Hat's statement when acquiring Ansible still ring true.

Before we look at rolling our sleeves up and installing Ansible, which we will do later in this chapter, we should look at some of its core concepts.

# Ansible versus other tools

If you look at the design principles in the first commit compared to the current version, you will notice that while there have been some additions and tweaks, the core principles remain pretty much intact:

- **Agentless**: Everything should be managed by the SSH daemon using the WinRM protocol in the case of Windows machines or API calls—there should be no reliance on custom agents or additional ports that need to be opened or interacted with on the target host. The machine running Ansible should need line of sight of the target resource network-wise.

- **Minimal**: You should be able to manage new remote machines without installing any new software on the target host; each Linux target host will typically have at least SSH and Python installed as part of a minimal installation, which is all needed to run Ansible.

- **Descriptive**: You should be able to describe your infrastructure, stack, or task in a language readable by machines and humans.

- **Simple**: The setup processes and the learning curve should be simple and intuitive.

- **Easy to use**: It should be the most accessible IT automation system ever.

A few of these principles make Ansible quite different from other tools. Let's examine the fundamental difference between Ansible and other tools, such as Puppet and **Chef**.

## Declarative versus imperative

When I started using Ansible, I had already implemented Puppet to help manage the stacks on the machines I was managing. As the configuration became increasingly complex, the Puppet code became highly complicated. This was when I started looking at alternatives, and some fixed some of the issues I was facing.

Puppet uses a custom declarative language to describe the configuration. Puppet then packages this configuration as a manifest that the agent running on each server then applies.

Using declarative language means that Puppet, Chef, and other configuration tools, such as **CFEngine**, all operate using the principle of eventual consistency, meaning that eventually, after a few runs of the agent, your desired configuration would be in place.

On the other hand, Ansible is an imperative language that, rather than just defining the end state of your desired outcome and letting the tool decide how it should get there, you also define the order in which tasks are executed to reach the state you have defined.

The example I use is as follows. We have a configuration where the following states need to be applied to a server:

1.  Create a group called Team.

2.  Create a user Alice and add her to the group Team.

3.  Create a user Bob, and add him to the group Team.

4.  Give the user Alice escalated privileges.

This may seem simple; however, when you execute these tasks using declarative language, you may, for example, find that the following happens:

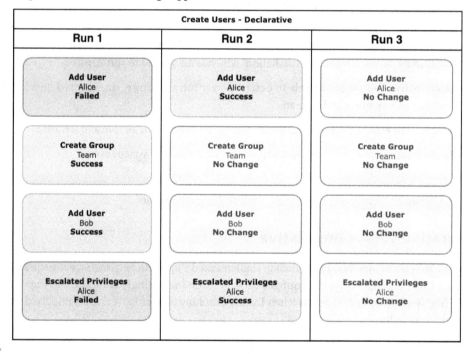

Figure 1.1 – An overview of what happens with the declarative run

So, what has happened here? Our tool has executed the tasks during **run 1** in the order of *2, 1, 3,* and *4*, meaning that the user Alice could not be created when the first task ran because the group Team did not exist.

However, as the group Team was created before the user Bob was created, Bob's user was created without any errors, and the final task, adding escalated privileges to the user Alice, failed because no user called Alice existed on the system for the escalated privileges to be applied to.

During **run 2**, the tasks were executed in the same order as **run 1**, but this time as a group called Team existed, the user Alice was created, and because Alice was present, that user was given escalated privileges.

No changes were needed during **run 3** as everything was as expected; that is, consistent.

Each subsequent run would continue until there was either a change to the configuration or on the host itself, for example, if Bob had annoyed Alice and she used her escalated privileges to remove the user Bob from the host. When the agent subsequently runs, Bob will be recreated as that is still our desired configuration, no matter what access Alice thinks Bob should have.

If we were to run the same tasks using an imperative language, then the following should happen:

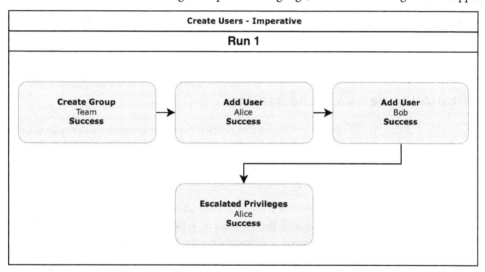

Figure 1.2 – An overview of what happens with the imperative run

The tasks are executed in the order we defined them, meaning that the Team group is created, the Alice and Bob users are added, and the escalated privileges to the Alice user are applied.

As you can see, both ways get to our final configuration and enforce our desired state. With the tools that use declarative language, it is possible to declare dependencies, meaning we can engineer the issue we encountered when running the tasks.

However, this example only has four steps; what happens when you have a few hundred steps that are launching servers in public cloud platforms and then installing software that needs several prerequisites?

This is the position I found myself in before I started to use Ansible. Puppet was great at enforcing my desired end configuration; however, to get there, I had to worry about building so much logic into my manifests to arrive at my desired state. In Puppet, this logic was using a function that allowed me as the end user to define my dependencies.

In the example we used, I would have had to define that users could only be created once the block of code that created the group had been run and the resource was present.

The more complex my code got, the more I fought the way the declarative tools wanted to run and the longer each execution would take because the tool had to consider my logic, which was a little hit and miss.

This became more annoying, as each successful run was getting close to taking about 40 minutes to complete. If I had dependency issues, I had to start from scratch with each failure and change to ensure that I was fixing the problem and not because things were starting to become consistent, so that typically meant having to redeploy a resource rather than running subsequent runs of my code. This made development very time-consuming, especially when it came to debugging the code, which sometimes involved trial and error.

This is not a great position to find yourself in when you are on the clock and must meet customer deadlines.

## Configuration versus orchestration

Another critical difference between Ansible and the other tools it is commonly compared to is that most of these tools have their origins as systems designed to deploy and police a configuration state.

They typically require an agent to be installed on each host; that agent discovers some information about the host it is installed on, then calls back to a central server saying, "*Hi, I am server XYZ. Could I please have my configuration?*" The server then decides what the configuration for the server looks like and sends it across to the agent, which then applies it. Typically, this exchange takes place every 15 to 30 minutes—this is great if you need to enforce a configuration on a server.

However, the way that Ansible has been designed to run allows it to act as an orchestration tool; for example, you can run it to launch a server in your VMware environment, and once the server has been launched, it can then connect to your newly launched machine and install a LAMP stack. Then, it never has to connect to that host again, meaning that all we are left with is the server, the LAMP stack, and nothing else, other than maybe a few comments in files to say that Ansible added some lines of configuration, which should be the only sign that Ansible was used to configure the host.

## Looking at some code

Before we finish this part of the chapter and move on to installing Ansible, let's quickly look at examples of some actual code. The following bash script installs several RPMs using the yum package manager:

```sh
#!/bin/sh
LIST_OF_APPS="dstat lsof mailx rsync tree vim-enhanced git whois"
yum install -y $LIST_OF_APPS
```

The following is a Puppet class that does the same task as the previous bash script:

```
class common::apps {
  package {
    [
      'dstat',
      'lsof',
      'mailx',
      'rsync',
      'tree',
      'vim-enhanced',
      'git',
      'whois',
    ]:
    ensure => installed,
  }
}
```

Next up, we have the same task using **SaltStack**:

```
common.packages:
  pkg.installed:
    - pkgs:
      - dstat
      - lsof
      - mailx
      - rsync
      - tree
      - vim-enhanced
      - git
      - whois
```

Finally, we have the same task again, this time using Ansible:

```
- name: "Install packages we need"
  ansible.builtin.yum:
```

```
name:
  - "dstat"
  - "lsof"
  - "mailx"
  - "rsync"
  - "tree"
  - "vim-enhanced"
  - "git"
  - "whois"
  - "iptables-services"
state: "present"
```

Even without going into any detail, you should be able to get the general gist of what each of the three examples is doing. While not strictly infrastructure, all three are valid examples of infrastructure as code.

This is where you manage the code that manages your infrastructure in precisely the same way as a developer would manage the source code for their application. You use source control, store it in a centrally available repository where you can collaborate with your peers, branch and use pull requests to check in your changes, and, where possible, write and execute unit tests to ensure that changes to your infrastructure are successful and error-free before deploying to production. This should be as automated as possible. Any manual intervention in the tasks mentioned could be a point of failure, and you should work to automate the task.

This approach to infrastructure management has a few advantages, one being that you, as system administrators, are using the same processes and tooling as your developer colleagues, meaning that any procedures that apply to them also apply to you. This makes for a more consistent working experience and exposes you to tools you may have yet to be exposed to or use.

Secondly, and more importantly, it allows you to share your work. Before this approach, this type of work seemed to others a dark art performed only by system administrators. Doing this work in the open allows you to have your peers review and comment on your configuration and do the same yourself to theirs. Also, you can share your work so that others can incorporate elements into their projects.

## My story: part two

Before we finish this part of the chapter, I would like to finish the story of my journey. As mentioned earlier in the chapter, I moved from my collection of scripts and runbooks to Puppet, which was great until my requirements moved away from managing just server configuration and maintaining the servers' state.

I needed to start managing infrastructure in public clouds. This requirement quickly started to frustrate me when using Puppet. At the time, Puppet's coverage of the APIs I needed to use for my infrastructure needed to be improved. I am assured it is a lot better now, but also, I found myself having to build too much logic into my manifests about the order in which each task was executed.

It was around this time, which was December 2014, that I decided to look at Ansible. I know the date because I wrote a blog post entitled *First Steps With Ansible*; I don't think I have looked back since. I have since introduced several of my work colleagues and customers to Ansible and have written books on the subject, including the first edition of the title you are reading now.

So far in this chapter, we have looked at my personal history with both Ansible and some of the other tools that Ansible is compared to, and we have discussed the differences between these tools and where Ansible originated.

Now we are going to start your journey with Ansible by looking at installing it and running our first Ansible playbook against a local virtual machine.

# Installing and running Ansible

Let's dive straight in and install Ansible. Throughout this book, I will assume you are running a macOS host machine or a Linux machine with an Ubuntu LTS release. While we will cover running Ansible on Windows 11 using the Linux subsystem for Windows, this book will not support using Windows as a host machine.

## Installing on macOS

You can install Ansible on your macOS host machine in a few different ways. I will cover both here. As we are discussing two different ways of installing Ansible, I recommend reading through this section and the *Pros and cons* subsection before choosing which installation method to use on your local machine.

### Installing with Homebrew

The first installation method is to use a package manager called **Homebrew**.

> **Note**
>
> Homebrew is a package manager of macOS. It can be used to install command-line tools and desktop packages. It describes itself as *"The missing package manager for macOS"*, and it usually is one of the first tools I install after a clean installation or when getting a new computer.

To install Ansible using Homebrew, you first need to install Homebrew. To do this, run the following command:

```
$ /usr/bin/ruby -e "$(curl -fsSL https://raw.githubusercontent.com/
Homebrew/install/master/install)"
```

At each step of the installation process, the installer will tell you exactly what it is going to do and also prompt you for any additional information it needs from you to complete the installation.

Once installed, or if you already have Homebrew installed, run the following commands to update your list of packages; if there are any updates, then you can also upgrade them:

```
$ brew update
$ brew upgrade
```

Finally, depending on how new your installation is or when you last used it, you might want to run the following command to check that your Homebrew installation is optimal:

```
$ brew doctor
```

Now that we have Homebrew installed, updated, and ready to `brew`, we can run the following to check what Ansible packages Homebrew has by running:

```
$ brew search ansible
```

As you can see from the results in the following screenshot, there are several packages returned in the search:

Figure 1.3 – Searching for Ansible using the brew command

We want the Ansible package; you can find out more about the package by running the following command:

```
$ brew info ansible
```

You can see the results of the command in the following screenshot:

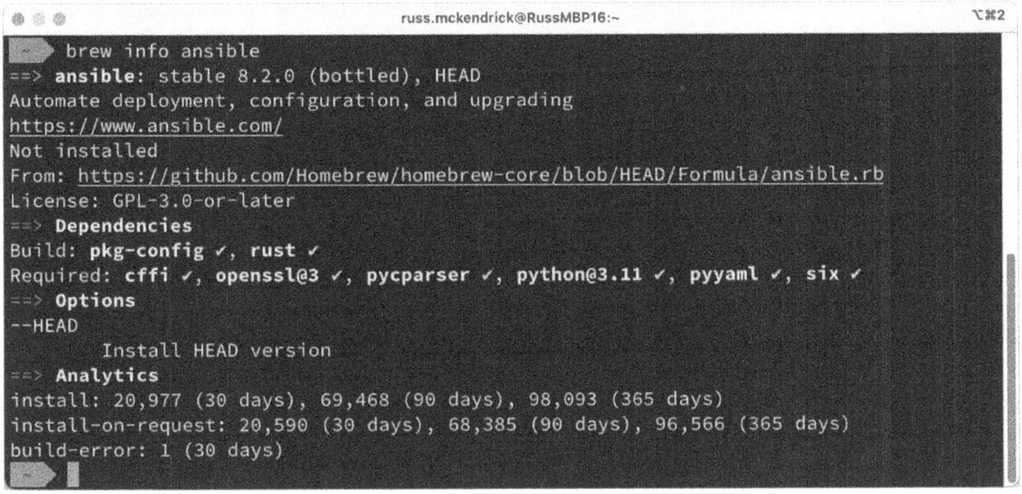

```
brew info ansible
==> ansible: stable 8.2.0 (bottled), HEAD
Automate deployment, configuration, and upgrading
https://www.ansible.com/
Not installed
From: https://github.com/Homebrew/homebrew-core/blob/HEAD/Formula/ansible.rb
License: GPL-3.0-or-later
==> Dependencies
Build: pkg-config ✓, rust ✓
Required: cffi ✓, openssl@3 ✓, pycparser ✓, python@3.11 ✓, pyyaml ✓, six ✓
==> Options
--HEAD
        Install HEAD version
==> Analytics
install: 20,977 (30 days), 69,468 (90 days), 98,093 (365 days)
install-on-request: 20,590 (30 days), 68,385 (90 days), 96,566 (365 days)
build-error: 1 (30 days)
```

Figure 1.4 – Viewing information on the Ansible package we will install

As you can see, the command returns information on the version of the package that will be installed along with a complete list of its dependencies; in the preceding screen, all of the dependencies have green ticks next to them because I already have them installed—yours may look different.

It also gives the URL to the Homebrew formula, which will be used to install the package. In our case, you can view formula details at `https://github.com/ Homebrew/homebrew-core/ blob/master/Formula/ansible.rb`.

To install Ansible using Homebrew, we must run the following command:

```
$ brew install ansible
```

This will download and install all dependencies and then the Ansible package itself.

Depending on how many dependencies are installed on your machine, this may take a few minutes.

Once installed, you should see something like the following screenshot:

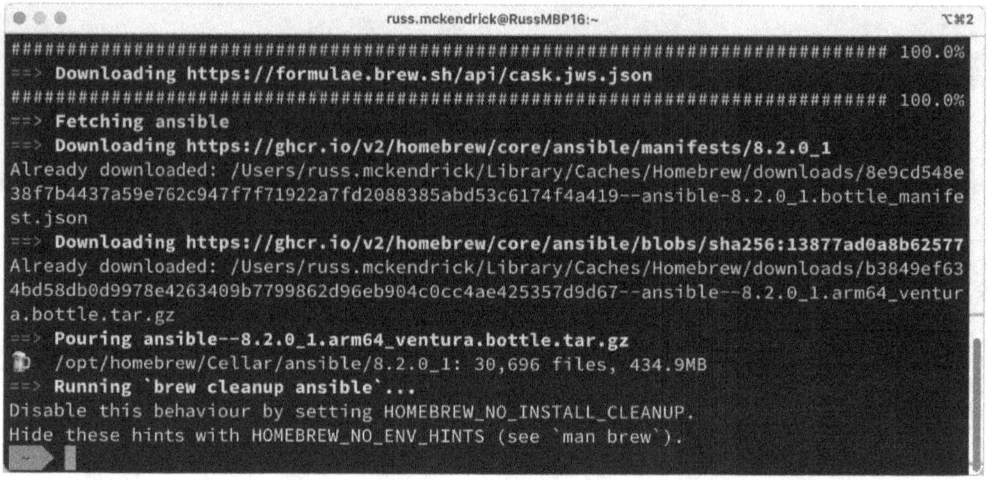

Figure 1.5 – Installing Ansible using Homebrew

As you can see from the preceding screenshot, Homebrew is quite verbose in its output, giving you both feedback on what it is doing and details on how to use the packages it installs.

The second of two installation methods we will look at for macOS is a more traditional one.

### Installing using pip

The second method, pip, is a more traditional approach to installing and configuring a Python package.

> **Note**
>
> pip is a package manager for Python software. It is a recursive acronym for **pip install packages**. It is a good frontend for installing packages from the **Python Package Index (PyPI)**.

Most modern macOS installations come with pip installed by default; depending on what you have installed on your machine, you may have to check which pip binary you have installed.

To do this, run the following commands:

```
$ pip --version
$ pip3 --version
```

One or both should return a version number and give you the path to the pip binary.

Depending on the version of pip you have installed, you may need to amend the following pip command, which is what we need to run to install Ansible:

```
$ pip install ansible
```

This command will download and install all the prerequisites to run Ansible on your system. While it is as verbose as Homebrew, its output contains information on what it has done rather than hints on what to do next:

```
russ.mckendrick@RussMBP16:~                                          ⌥⌘2
ges (from ansible-core~=2.15.2->ansible) (40.0.2)
Requirement already satisfied: packaging in /opt/homebrew/lib/python3.11/site-packages
 (from ansible-core~=2.15.2->ansible) (23.1)
Collecting resolvelib<1.1.0,>=0.5.3 (from ansible-core~=2.15.2->ansible)
  Downloading resolvelib-1.0.1-py2.py3-none-any.whl (17 kB)
Requirement already satisfied: MarkupSafe>=2.0 in /opt/homebrew/lib/python3.11/site-pa
ckages (from jinja2>=3.0.0->ansible-core~=2.15.2->ansible) (2.1.2)
Requirement already satisfied: cffi>=1.12 in /opt/homebrew/lib/python3.11/site-package
s (from cryptography->ansible-core~=2.15.2->ansible) (1.15.1)
Requirement already satisfied: pycparser in /opt/homebrew/lib/python3.11/site-packages
 (from cffi>=1.12->cryptography->ansible-core~=2.15.2->ansible) (2.21)
Downloading ansible-8.2.0-py3-none-any.whl (45.1 MB)
━━━━━━━━━━━━━━━━━━━━━━━━━━━ 45.1/45.1 MB 77.9 MB/s eta 0:00:00
Downloading ansible_core-2.15.2-py3-none-any.whl (2.2 MB)
━━━━━━━━━━━━━━━━━━━━━━━━━━━ 2.2/2.2 MB 82.8 MB/s eta 0:00:00
Installing collected packages: resolvelib, ansible-core, ansible
Successfully installed ansible-8.2.0 ansible-core-2.15.2 resolvelib-1.0.1
```

Figure 1.6 – Installing Ansible using Pip

As you can see from the small amount of output, many of the requirements were already satisfied.

### Pros and cons

So, now that we have covered some of the different ways of installing Ansible on macOS, which is best? Well, there is no real answer to this as it comes down to personal preference. Both methods will install the latest versions of Ansible. However, Homebrew tends to be a week or two behind the current release.

If you have a lot of packages already installed using Homebrew, then you will be used to running the following commands:

```
$ brew update
$ brew upgrade
```

Just run these occasionally to update your installed packages to the latest versions. If you already do this, it makes sense to use Homebrew to manage your Ansible installation.

If you are not a Homebrew user and want to ensure that you immediately have the latest version installed, use the pip command to install Ansible. Upgrading to the latest version of Ansible is as simple as running the following command:

```
$ pip install ansible --upgrade
```

Should you need to, you can install older versions of Ansible using Homebrew and pip.

To do this using Homebrew, you need to remove the current version by running the following command:

```
$ brew uninstall ansible
```

Then, you can install an earlier version of the package by running the following command:

```
$ brew install ansible@2.0
```

While this will install an earlier version, you do not have much choice in which version you get. If you really need an exact version, you can use the pip command to install it. For example, to install Ansible 2.3.1.0, you would need to run:

```
$ pip install ansible==2.3.1.0
```

It is essential to note that you should never need to do this, and I do not recommend it.

However, I have found that on rare occasions, I have had to downgrade to help debug quirks in my playbooks introduced by upgrading to a later version of Ansible for playbooks that I last touched a few years ago.

As mentioned, I spend the bulk of my time in front of a macOS machine both during my day job and at home, so which of the two methods do I use?

Primarily, I use Homebrew as I have several other tools installed using Homebrew. However, if I need to roll back to a previous version, I use pip and then return to Homebrew once the issue is resolved.

## Installing on Linux

There are a few different ways of installing Ansible on Ubuntu. However, I am only going to cover one of them here. While there are packages available for Ubuntu that you can install with apt, they tend to become out of date quickly and are typically behind the current release.

If you wish to install using the apt package manager, then you can run the following command:

```
$ apt install ansible
```

> **Note**
>
> **Advanced Packaging Tool (APT)** is the package manager that ships with Debian-based systems, including Ubuntu. It is used to manage `.deb` files.

Because of this, we will be using pip. The first thing to do is install pip, and this can be done by running the following commands:

```
$ sudo -H apt-get update
$ sudo -H apt-get install python3-pip
```

The first of the two `apt-get` commands downloads all the update files, ensuring that the package list is up to date on your Ubuntu installation, and the second command installs the `python3-pip` package and its dependencies.

Once pip is installed, the instruction for installing Ansible is similar to installing on macOS. Run the following command:

```
$ sudo -H pip install ansible
```

This will download and then install Ansible and its requirements, as shown in the following screenshot:

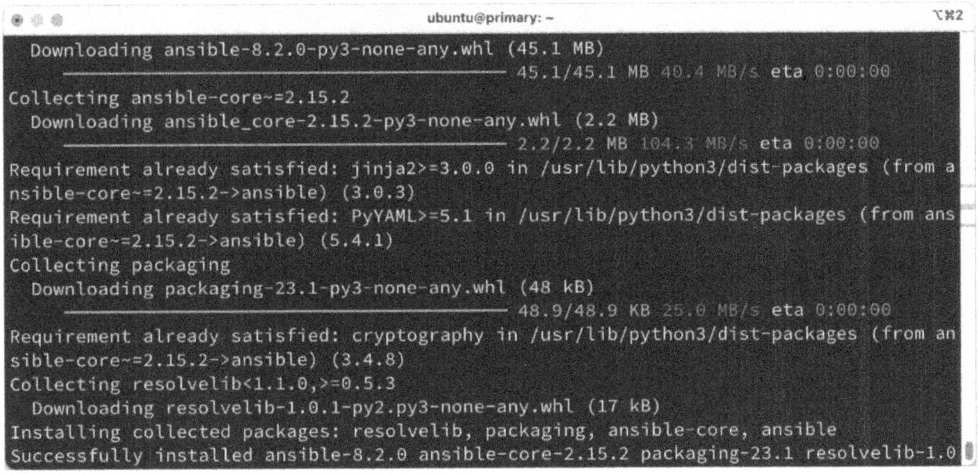

Figure 1.7 – Installing Ansible using pip on Ubuntu

Once installed, you can upgrade it by using the following command:

```
$ sudo -H pip install ansible --upgrade
```

Also, downgrading Ansible uses the same command:

```
$ sudo -H pip install ansible==2.3.1.0
```

The preceding commands should work on most Linux distributions, such as Rocky Linux, Red Hat Enterprise Linux, Debian, and Linux Mint, to name a few.

A lot of these distributions have their own package managers you can also use to install Ansible; for example, on Red Hat-based distributions such as Red Hat Enterprise Linux or Rocky Linux, you could also run:

```
$ dnf install ansible-core
```

Please consult the document for more details on installing whatever your Linux distribution of choice is.

## Installing on Windows 11

The last platform we will cover is Windows 11; well, sort of. While it is technically possible to run Ansible on a Windows 11 natively, it is not something I would recommend attempting as it is one of those tasks where the phrase "*just because you can doesn't mean you should*" applies, as getting all of the dependencies installed and configured just right can be very troublesome, and maintaining them is more so.

Luckily, Microsoft—and as a long-term Linux System administrator typing this, this still feels strange—has excellent native support for running Linux systems seamlessly within Windows 11.

Open the Microsoft Store and search for Ubuntu; you should see something like the following screen:

Figure 1.8 – Finding Ubuntu in the Microsoft Store

Click the **Get** button to download Ubuntu. Once downloaded, we have Ubuntu on our Windows 11 host, but we still need something to run it on. To run it, we need to enable Windows Subsystem for Linux.

To enable this, open a PowerShell window by typing **PowerShell** into your Windows Search bar and opening the **Windows PowerShell** application; once you are at a terminal prompt, run the following command:

```
$ wsl --install
```

Follow the onscreen prompts, and once everything is installed, restart your Windows 11 host.

Once rebooted, you should have something like the following prompt pop-up once you log back in:

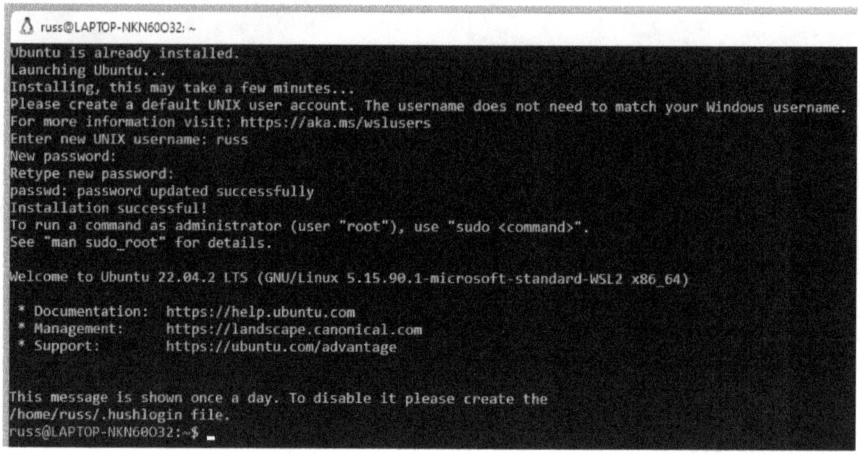

Figure 1.9 – Completing the Ubuntu installation on Windows 11

Once the installation is completed, I like to switch out of the default Windows Subsystem for the Linux terminal and use the Microsoft Terminal, which you can grab for free in the Microsoft Store.

Once you have your preferred terminal emulator open and you are sitting at a prompt in your Windows Subsystem for Linux Ubuntu installation, you can run the same commands we ran to install Ansible on Linux, which are as follows:

```
$ sudo -H apt-get update
$ sudo -H apt-get install python3-pip
$ sudo -H pip install ansible
```

Once you have run these commands, you should see an output that looks like the following screenshot:

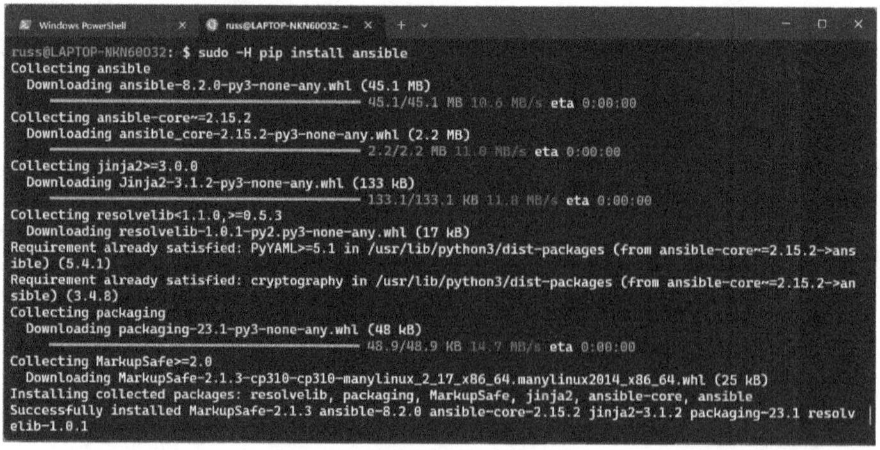

Figure 1.10 – Installing Ansible in Ubuntu on Windows 11

As you can see, everything works as if you were running an Ubuntu machine, allowing you to run and maintain your Ansible installation in precisely the same way.

> **Note**
>
> The **Windows Subsystem for Linux** (**WSL**) is not running on a virtual machine. It is a full native Linux experience baked right into Windows 11. It targets developers who must run Linux tools as part of their toolchain. While the overall support for Linux commands is excellent, I recommend reading through the FAQs written and maintained by Microsoft to get an idea of the limits and quirks of the subsystem. The FAQ can be found at `https://learn.microsoft.com/en-us/windows/wsl/faq`.

As mentioned, while this is a viable way of running an Ansible control node on a Windows-based machine, some of the other tools we will cover in future chapters may not work with Windows. So, while you may follow along using the Ubuntu instructions, some parts may not work.

## Launching a virtual machine

To launch a virtual machine to run our first set of Ansible commands against, we will use a tool called **Multipass**. This tool allows you to run Ubuntu virtual machines on your local host. It works with macOS, Linux, and Windows.

To install Multipass on macOS, we can use Homebrew and run the following command:

```
$ brew install multipass
```

To install on Ubuntu, you can run the following command:

```
$ snap install multipass
```

Finally, for Windows 11 users, you will have to first download and install the VirtualBox Windows executable from `https://www.virtualbox.org/wiki/Downloads` and then download and install Multipass from https://multipass.run/install. I recommend reading through the installation notes for Windows, which can be found at the following URL https://multipass.run/docs/installing-on-windows before installing.

> **Note**
>
> While you can run the same commands in Ubuntu running under Windows Subsystem for Linux, you will need to replace all references of the `multipass` command with `multipass.exe` so that the Windows version of Multipass is called.

Next, check out the GitHub repo accompanying this title and open your terminal in the `Chapter01` folder—if you are running Windows 11, you must open an Ubuntu terminal, not a Windows one.

> **Important**
>
> Before we start, a quick word of warning: the folder Chapter01 contains an OpenSSH key-pair, which will be used to access the local machine. It is important that you do not use this key-pair anywhere other than this example on your local machine as the key-pair is publicly accessible, which is not considered secure.

You will see several files in the Chapter01 folder. The one we are going to use when launching the virtual machine is called **cloud-init.yaml**. This file contains some back configuration to add a user called vmadmin and attach the public portion of an OpenSSH key to the user, meaning that when executing Ansible, we can use the private part of the OpenSSH key to authenticate as the vmadmin user.

The command we are going to run to launch the virtual machine, which will be called **ansiblevm**, is as follows:

```
$ multipass launch -n ansiblevm --cloud-init cloud-init.yaml
```

Once the virtual machine has been launched, which may take a short while when you first run the command as it will download a virtual machine image, you need to run the following command to get some information on the newly created ansiblevm virtual machine:

```
$ multipass info ansiblevm
```

The following screen shows me starting and viewing the information on the virtual machine:

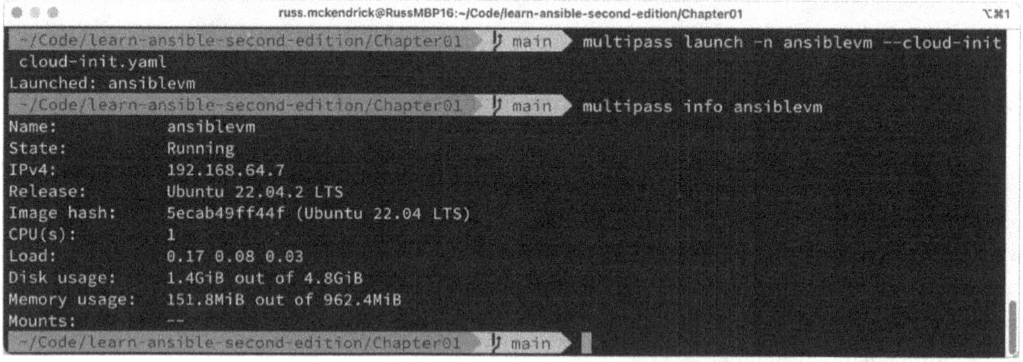

Figure 1.11 – Launching our virtual machine

Now that we have our virtual machine running and we have checked out the basic information, you will need to note the IP address, which in my case is 192.168.64.7. The IP address will be different when you launch the virtual machine on your host.

Before we run our first Ansible playbook, you must make a copy of the `hosts-simple.example` and `hosts.example` files and remove the `.example` in the filename by running the following commands:

```
$ cp -pr hosts-simple.example hosts-simple
$ cp -pr hosts.example hosts
```

Once you have made a copy of the file, open the newly created files and replace just the text that says `paste_your_ip_here` with the IP address of the `ansiblevm` virtual machine; in my case, the `hosts-simple` file went from:

```
paste_your_ip_here.nip.io ansible_user=vmadmin ansible_private_key_
file=./example_key
```

to reading:

```
192.168.64.7.nip.io ansible_user=vmadmin ansible_private_key_file=./
example_key
```

Once you have changed both the `hosts-simple` and `hosts` files, you are ready to run your first Ansible Playbook.

# An introduction to playbooks

Typically, in IT, a **playbook** is a set of instructions run by someone when something happens; a little vague, I know, but stay with me. These range from building and configuring new server instances to deploying code updates and dealing with problems when they occur.

In the traditional sense, a playbook is typically a collection of scripts or instructions for a user to follow, and while they are meant to introduce consistency and conformity across systems, even with the best intentions, this is seldom the case.

This is where Ansible comes in. Using an Ansible playbook, you are telling it to apply these changes and commands against these sets of hosts rather than having to log in and start working your way through the playbook manually.

Before we run a playbook, let's discuss how we provide Ansible with a list of hosts to target. To do this, we will be using the `ansible.builtin.setup` module. This connects to a host and then fetches as much information on the host as possible.

## Host inventories

To provide a list of hosts, we need to provide an inventory list. This is in the form of a host's file.

In its simplest form, our host's file could contain a single line just like our `hosts-simple` file:

```
192.168.64.7.nip.io ansible_user=vmadmin ansible_private_key_file=./
example_key
```

This tells Ansible that the host we want to contact is `192.168.64.7.nip.io` (please remember your IP address will be different) using the username `vmadmin`. If we didn't provide the username, it would fall back to the user you are logged into your Ansible control host as, which in my case is the user `russ`, which does not exist on the `ansiblevm` we launched. The final part tells Ansible to use the private OpenSSH key file called `example_key`, which we installed the public portion of to the `vmadmin` user when we launched the virtual machine.

> **Note**
>
> We are using `https://nip.io`, a free service that provides free wildcard DNS entries for any hostname containing an IP address. This means that our domain `192.168.64.7.nip.io` will resolve to `192.168.64.7` when a DNS lookup is made against the domain.

To run the `ansible.builtin.setup` module, we need to run the following command from within the `Chapter01` folder where your updated `hosts-simple` and `example_key` files are stored, making sure to update the IP address to your own:

```
$ ansible -i hosts-simple 192.168.64.7.nip.io -m ansible.builtin.setup
```

If everything works as expected, you should see a lot of output, which specifies some quite detailed and low-level information about your host. You should see something like the following:

Figure 1.12 – The start of the output of me running the ansible.builtin.setup module

As you can see from the preceding screenshot, Ansible has quickly found out a lot of information on our Vagrant box. The screenshot shows the IP addresses configured on the machine, along with the IPv6 addresses. It has recorded the time and date, and if you scroll through your output, you will see a lot of information returned detailing the host.

Let's go back to the command we ran:

```
$ ansible -i hosts-simple 192.168.64.7.nip.io -m ansible.builtin.setup
```

As you can see, we are loading the hosts-simple file using the -i flag. We could have also used --inventory=hosts-simple, which loads our inventory file. The next part of the command is the host to the target. In our case, this is 192.168.50.4.nip.io. The final part of the command, -m, tells Ansible to use the setup module. We could have also used --module-name= ansible. builtin.setup.

This means that the full command if we didn't use shorthand would be:

```
$ ansible --inventory=hosts-simple simple 192.168.64.7.nip.io
--module-name=ansible.builtin.setup
```

As already mentioned, the hosts-simple file is as basic as we can get it. The following is a more common host inventory file:

```
ansiblevm ansible_host=192.168.64.7.nip.io
[ansible_hosts]
ansiblevm
[ansible_hosts:vars]
ansible_connection=ssh
ansible_user=vmadmin
ansible_private_key_file=./example_key
host_key_checking=False
```

This is the content of the file called just hosts; as you can see, there is a lot more going on, so let's quickly work through it from top to bottom.

The first line defines our single host. Unlike the simple example, we will be calling our target host ansiblevm and grouping it together in a group called ansible_hosts, so we are giving Ansible details of where it can SSH to. This means we can now use the name ansiblevm when referring to 192.168.64.7.nip.io. This means our command would now look something like this:

```
$ ansible -i hosts ansiblevm -m ansible.builtin.setup
```

Next up in the file, we are creating a group of hosts called ansible_hosts and, in that group, we are adding our single host ansiblevm. This means that we can also run:

```
$ ansible -i hosts ansible_hosts -m ansible.builtin.setup
```

If we had more than just a single host in the group, the preceding command would have looped through all of them. The final section of the hosts file sets up some common configuration options for all of the hosts in the boxes group. In this case, we are telling Ansible that all of the hosts in the group are using SSH, the user is vmadmin, the private key at ./example_key should be used, and it should not check the host key when connecting.

We will be revisiting the inventory host files in later chapters. From now on, we will use the hosts file to target the ansible_hosts group.

## Playbooks

In the previous section, running the `ansible` command allowed us to call a single module.

In this section, we are going to look at calling several modules. The following playbook is called **playbook01.yml**. It calls the `ansible.builtin.setup` module we called in the previous section and then uses the `ansible.builtin.debug` module to print a message to the screen:

```
---
- name: "A simple playbook"
  hosts: ansible_hosts
  gather_facts: true
  become: true
  become_method: "ansible.builtin.sudo"
  tasks:
    - name: "Output some information on our host"
      ansible.builtin.debug:
        msg: "I am connecting to {{ ansible_nodename }} which is
running {{ ansible_distribution }} {{ ansible_distribution_version }}"
```

Before we break the configuration down, let's look at the results of running the playbook. To do this, use the following command:

```
$ ansible-playbook -i hosts playbook01.yml
```

This will connect to our host, gather information on the system, and then return just the information we want in a message:

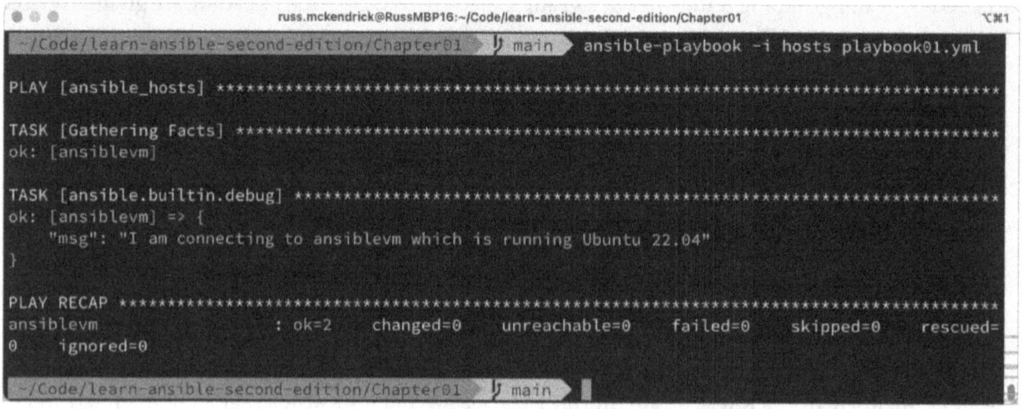

Figure 1.13 – The output of running ansible-playbook01.yml

The first thing you will notice about the playbook is that it is written in **YAML**, a recursive acronym for **YAML Ain't Markup Language**. YAML was designed to be a human-readable data serialization standard that all programming languages can use. It is commonly used to help define configurations.

The indentation is very important in YAML as it is used to nest and define areas of the file. Let's look at our playbook in more detail:

```
---
```

While these lines might not seem like much, they are used as document separators, as Ansible compiles all the YAML files into a single file. It is essential for Ansible to know where one document ends and another begins.

Next up, we have the configuration for the playbook. As you can see, this is where the indentation starts to come into play:

```
- name: "A simple playbook"
  hosts: ansible_hosts
  gather_facts: true
  become: true
  become_method: "ansible.builtin.sudo"
```

The - tells Ansible that this is the start of a section. From there, key-value pairs are used. These are as follows:

- name: This gives a name to the playbook run.
- hosts: This tells Ansible the host or host group to target in the playbook. This must be defined in a host inventory like the ones we covered in the previous section.
- gather_facts: This tells Ansible to run the ansible.builtin.setup module when it first connects to the host. This information is then available to the playbook during the run.
- become: This is present because we are connecting to our host as a basic user, in this case, the **vmadmin** user. Ansible may not have enough access privileges to execute some of the commands we are telling it to, so this instructs Ansible to execute all of its commands as the root user.
- become_method: This tells Ansible how to become the root user; in our case, we have a passwordless sudo configured by the cloud-init script we ran when launching the virtual machine, so we are using ansible.builtin.sudo.
- tasks: These are the tasks we can tell Ansible to run when connected to the target host.

You will notice that from here, we move the indentation across again. This defines another section of the configuration. This time it is for the tasks:

```
- name: "Output some information on our host"
  ansible.builtin.debug:
```

```
        msg: "I am connecting to {{ ansible_nodename }} which is
running {{ ansible_distribution }} {{ ansible_distribution_version }}"
```

As we have already seen, the only task we run is the `ansible.builtin.debug` module. This module allows us to display output in the Ansible playbook run stream you saw when we ran the playbook.

You may have already noticed that the information between the curly brackets is made up of the keys from the `ansible.builtin.setup` module. Here, we are telling Ansible to substitute the value of each key wherever we use the key. We will be using this a lot in our playbooks. We will also be defining our own key values to use as part of our playbook runs.

Let's extend our playbook by adding another task. The following can be found as **playbook02.yml**:

```
---
- name: "Update all packages"
  hosts: "ansible_hosts"
  gather_facts: true
  become: true
  become_method: "ansible.builtin.sudo"
  tasks:
    - name: "Output some information on our host"
      ansible.builtin.debug:
        msg: "I am connecting to {{ ansible_nodename }} which is
running {{ ansible_distribution }} {{ ansible_distribution_version }}"
    - name: "Update all packages to the latest version"
      ansible.builtin.apt:
        name: "*"
        state: "latest"
        update_cache: true
```

As you can see, we have added a second task that calls the `ansible.builtin.apt` module. This module is designed to help us interact with the package manager used by Ubuntu and other Debian-based operating systems called `apt`. We are setting three key values here:

- `name`: This is a wildcard. It tells Ansible to use all of the installed packages rather than just a single named package. For example, we could have used something such as `apache2` here to target Apache.

- `state`: Here, we are telling Ansible to ensure the package we have defined in the name key is the `latest` version. As we have named all of the installed packages, this will update everything we have installed.

- `update_cache`: As the virtual machine image we downloaded was optimized for being small, it does not contain any information on the available package; by setting `update_cache` to `true`, this will download a list of all package and version information.

Run the playbook using the following command:

```
$ ansible-playbook -i hosts playbook02.yml
```

This will give us the following results:

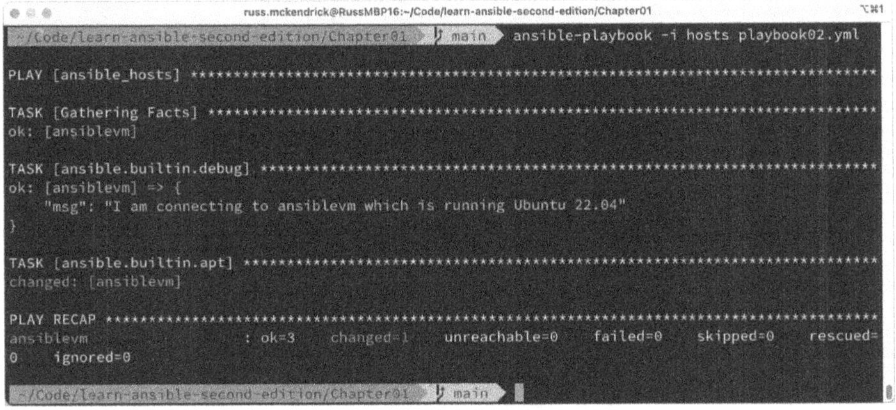

Figure 1.14 – The output of running ansible-playbook02.yml

The `ansible.builtin.apt` task has been marked as changed on the host box. This means that packages were updated.

Rerunning the same command shows the following results:

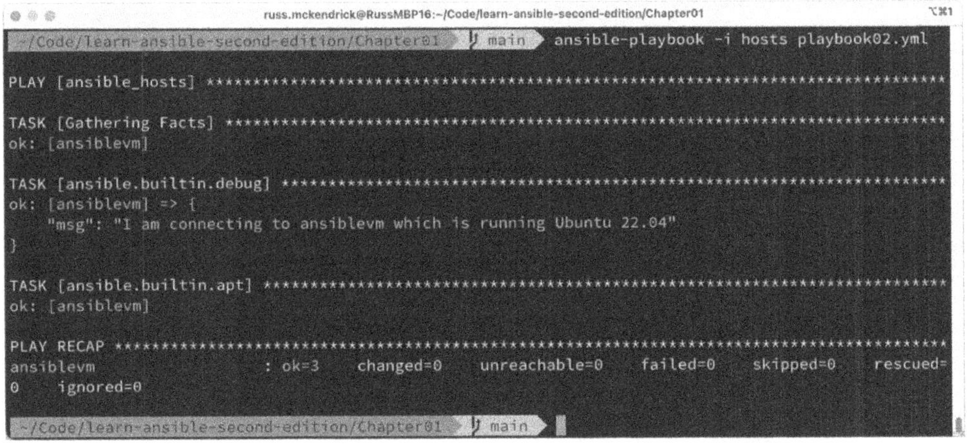

Figure 1.15 – The output of rerunning ansible-playbook02.yml

As you can see, the `ansible.builtin.apt` task is now showing as `ok` on our host. This is because there are currently no longer any packages requiring updates.

Before we finish this quick look at playbooks, let's do something more interesting.

The playbook, `playbook03.yml`, adds NTP installing, configuring, and starting capabilities to our virtual machine. It also uses a template to add a custom NTP config file to our virtual machine.

The `vars` section allows us to configure our own key-value pairs. In this case, we are providing a list of NTP servers, which we will be using later in the playbook:

```
vars:
  ntp_servers:
    - "0.uk.pool.ntp.org"
    - "1.uk.pool.ntp.org"
    - "2.uk.pool.ntp.org"
    - "3.uk.pool.ntp.org"
```

We are actually providing four different values for the same key. These will be used in the template task. We could have also written this as follows:

```
vars:
  ntp_servers: [ "0.uk.pool.ntp.org", "1. uk.pool.ntp.org", "2.
uk.pool.ntp.org", "3. uk.pool.ntp.org" ]
```

However, this is a little more difficult to read. The next new section is `handlers`. A **handler** is a task that is assigned a name and called at the end of a playbook run depending on what tasks have changed:

```
handlers:
  - name: "Restart ntp"
    ansible.builtin.service:
      name: "ntp"
      state: "restarted"
```

In our case, the restart ntp handler uses the `ansible.builtin.service` module to restart ntp. Next up, we have two new tasks, starting with installing the NTP service and the `sntp` and `ntp-doc` packages using `ansible.builtin.apt`:

```
- name: "Install packages"
  ansible.builtin.apt:
    state: "present"
    pkg:
      - "ntp"
      - "sntp"
      - "ntp-doc"
```

As we need to install three packages, we need a way to provide three different package names to the `ansible.builtin.apt` module so that we don't have to have three different tasks for each of the package installations. To achieve this, we use the `pkg` option rather than the `name` option, where you can only define a single package to install. Rather than using `latest`, we are using `present`; this will mean that our packages don't get updated if they are already installed.

The final addition to the playbook is the following task:

```
- name: "Configure NTP"
  ansible.builtin.template:
    src: "./ntp.conf.j2"
    dest: "/etc/ntp.conf"
    mode: "0644"
  notify: "Restart ntp"
```

This task uses the `ansible.builtin.template` module. To read a template file from our Ansible controller, process it and upload the processed template to the host machine. Once uploaded, we are telling Ansible to notify the `restart ntp` handler if there have been any changes to the configuration file we are uploading.

In this case, the template file is the `ntp.conf.j2` file in the same folder as the playbooks, as defined in the `src` option. This file looks like this:

```
# {{ ansible_managed }}
driftfile /var/lib/ntp/drift
restrict default nomodify notrap nopeer noquery
restrict 127.0.0.1
restrict ::1
{% for item in ntp_servers %}
server {{ item }} iburst
{% endfor %}
includefile /etc/ntp/crypto/pw
keys /etc/ntp/keys
disable monitor
```

The bulk of the file is the standard NTP configuration file, with the addition of a few Ansible parts. The first addition is the very first line:

```
# {{ ansible_managed }}
```

If this line wasn't there every time we ran Ansible, the file would be uploaded, which would count as a change and the restart ntp handler would be called, meaning that even if there were no changes, NTP would be restarted.

The next part loops through the `ntp_servers` values we defined in the `vars` section of the playbook:

```
{% for item in ntp_servers %}
server {{ item }} iburst
{% endfor %}
```

For each of the values, add a line that contains the word `server`, the value or `{{ item }}`, and then `iburst`.

Now that we know what we have added to the playbook and have an idea of the additional tasks that will be performed, let's run it using the following command:

```
$ ansible-playbook -i hosts playbook03.yml
```

The following screen just shows the additional tasks and not the full output as we know it will be just marked as ok:

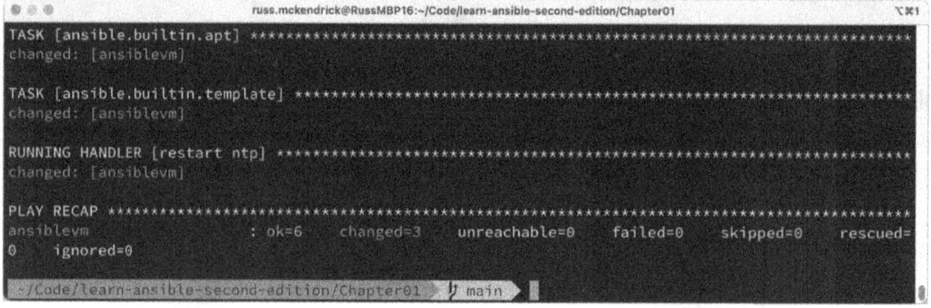

Figure 1.16 – The output of running ansible-playbook03.yml

This time, we have three changed tasks. Running the playbook again shows the following:

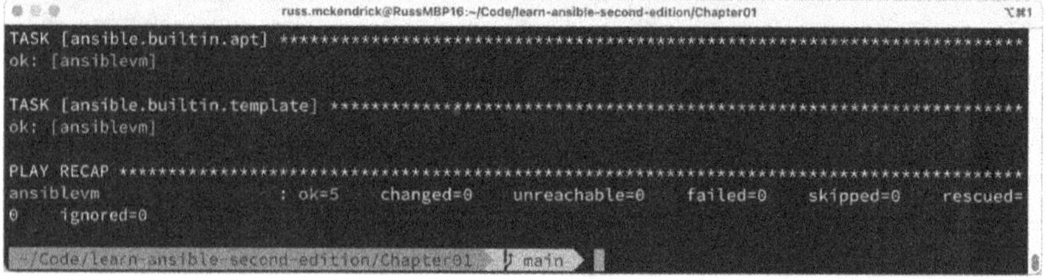

Figure 1.17 – The output of rerunning ansible-playbook03.yml

As expected, there are no changes because we haven't changed the playbook or anything on the virtual machine, and Ansible is reporting everything as ok. Also, because no changes were detected to the NTP config file, the Handler to restart NTP did not need to be called, and therefore it doesn't appear in the output.

Before we finish, let's launch a second virtual machine by running the following command:

```
$ multipass launch -n ansiblevm2 --cloud-init cloud-init.yaml
```

Once the second virtual machine has started, run the following command to get some information on the new virtual machine:

```
$ multipass info ansiblevm2
```

Now that we know the IP address, we can add two new lines to our `hosts` file. First of all, to define the new host, add the following code (updating it so it uses the correct IP address) underneath where our original host is defined:

```
ansiblevm2 ansible_host=192.168.64.8.nip.io
```

Then, add `ansiblevm2` to the `ansible_hosts` group:

```
[ansible_hosts]
ansiblevm
ansiblevm2
```

Then, rerun the playbook using the following:

```
$ ansible-playbook -i hosts playbook03.yml
```

As you can see, the same commands run, but now we are targeting both virtual machines, the original virtual machine has no changes, and all the changes are applied to the newly deployed host:

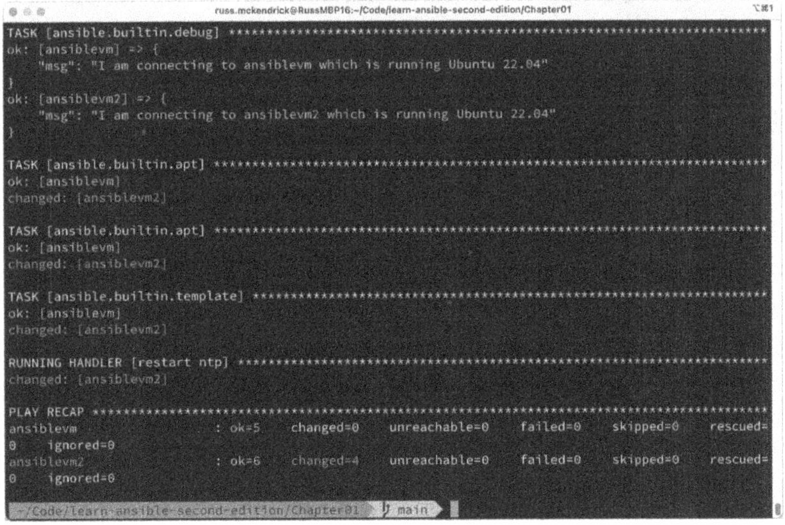

Figure 1.18 – The output of rerunning ansible-playbook03.yml against two virtual machines

If you rerun the command, you will see that everything is now shown as ok as there are no further changes.

Before we move on to the summary, let's tidy up our two virtual machines and remove them as we won't need them again. To do this, run the following command:

```
$ multipass delete --purge ansiblevm ansiblevm2
```

As I am sure you would have already guessed, this deletes the virtual machines and then purges the configuration and files.

# Summary

In this chapter, we have taken our first steps with Ansible by installing it locally and then, using Vagrant, launching a virtual machine to interact with. We learned about basic host inventory files and used the Ansible command to execute a single task against our virtual machine.

We then looked at playbooks, starting with a basic playbook that returned some information on our target before progressing to a playbook that updates all the installed operating system packages before installing and configuring the NTP service.

By the end of the chapter, we had launched a second virtual machine and quickly brought it up to the same configuration level as our first virtual machine.

In the next chapter, we will look at Ansible Galaxy and discuss how Ansible packages up and maintains its community modules.

# Further reading

In this chapter, we mentioned Puppet and SaltStack:

- **Puppet** is a configuration management tool that runs a server/agent configuration. It comes in two flavors—an open source version and an enterprise version that Puppet, the company, supports. It is a declarative system and is closely tied to Ruby. For more information on Puppet, see `https://www.puppet.com/`.

- **SaltStack** is another configuration management tool. It is highly scalable and, while it shares a design approach with Ansible, it works in a similar way to Puppet in that it has a server/agent approach. You can find more information on SaltStack at `https://www.vmware.com/support/acquisitions/saltstack.html`.

- I also mentioned my personal blog, which you can find at `https://www.russ.foo/`.

We used the following Ansible modules, and you can find out more information on each module at the following links:

- `ansible.builtin.setup`: `https://docs.ansible.com/ansible/latest/collections/ansible/builtin/setup_module.html`

- `ansible.builtin.debug`: `https://docs.ansible.com/ansible/latest/collections/ansible/builtin/debug_module.html`

- `ansible.builtin.apt`: `https://docs.ansible.com/ansible/latest/collections/ansible/builtin/apt_module.html`

- `ansible.builtin.template`: `https://docs.ansible.com/ansible/latest/collections/ansible/builtin/template_module.html`

- `ansible.builtin.service`: `https://docs.ansible.com/ansible/latest/collections/ansible/builtin/service_module.html`

# 2

# Exploring Ansible Galaxy

Welcome to our second chapter. Here, we are going to be looking at the `ansible-galaxy` command; we are going to be covering the features provided by the command and also discussing its importance in the development of Ansible over the last few years.

**Ansible Galaxy** is an online repository of community-contributed roles; we will discover some of the best roles available, how to use them, and how to create your own role and have it hosted on Ansible Galaxy.

By the end of the chapter, we will have worked through the following topics:

- The Ansible release life cycle

- Introduction to Ansible Galaxy

- What is a role?

- Publishing to and using Ansible Galaxy roles

- Ansible collections

- Ansible Galaxy commands

Before we start exploring Ansible Galaxy, let's discuss the Ansible core release life cycle and how it has changed over the last few years, as these changes have made it an essential component of the Ansible ecosystem.

## Technical requirements

In this chapter, we will again make use of **Multipass**, the tool that we covered in *Chapter 1, Installing and Running Ansible*, and the GitHub repository that accompanies this title, which can be found at `https://github.com/PacktPublishing/Learn-Ansible-Second-Edition`.

# The Ansible release life cycle

In the previous chapter, when we installed Ansible, the keen-eyed among you may have spotted a few different Ansible packages installed using the `sudo -H pip install ansible` command.

What follows is an edited version of the output of installing Ansible using `pip`:

```
$ sudo -H pip install ansible
Collecting ansible
  Downloading ansible-8.2.0-py3-none-any.whl (45.1 MB)
Collecting ansible-core~=2.15.2
  Downloading ansible_core-2.15.2-py3-none-any.whl (2.2 MB)
Installing collected packages: resolvelib, packaging, ansible-core,
ansible
Successfully installed ansible-8.2.0 ansible-core-2.15.2
packaging-23.1 resolvelib-1.0.1
```

As you can see, two main Ansible packages were installed: `ansible-8.2.0` and `ansible-core-2.15.2`. Before we discuss the difference between the two packages, let's quickly discuss how Ansible was maintained and packaged up until `version 2.9` of Ansible.

Every version of Ansible before `version 2.10` shipped with many modules baked into the release; while Ansible was new and its user base and functionality were focused on a few tasks, it was easy to manage and maintain the releases of these modules as part of the main Ansible code base, which was maintained by the Ansible team at the official GitHub repo, which you can find at `https://github.com/ansible/ansible`.

By the end of *Chapter 1, Installing and Running Ansible*, we had used a total of five modules, all of which are built into Ansible; these were the following:

- `ansible.builtin.setup`: A module that discovers information on the target host and makes it available during the playbook run

- `ansible.builtin.service`: A module that manages the state of a service on the target host

- `ansible.builtin.debug`: This module allows you to print statements during your playbook execution

- `ansible.builtin.apt`: This module manages packages on the target host using the apt package manager

- `ansible.builtin.template`: This module brings templating to Ansible, allowing you to output files to your target hosts

That was just a single playbook that did a single task; add to that the sheer number of modules to give you an idea of the numbers—there are currently 95 modules in the Amazon AWS namespace and 282 in the Microsoft Azure namespace—that's over 370 modules that cover the basic functionality of two different namespaces. There are, at the time of writing, over 40 different namespaces.

You might be asking, *"Wait, what is a namespace?"* The modules are now grouped into collections, and each collection has its namespace; some examples are the following:

- **ansible.builtin**: As you might have already guessed, the modules within this namespace provide some of Ansible's core functionality

- **amazon.aws**: These are the official Amazon Web Services modules

- **azure.azcollection**: Here, you will find the official Microsoft Azure modules

- **kubernetes.core**: If you want to work with Kubernetes, these will be your modules

Again, you might be thinking, *"That's useful information, but what has this got to do with anything?"*; well, Ansible used to have a few major releases a year and had to ship a release that had slowly grown to include thousands of modules, plus their associated plugins meant that the release process quickly became unmanageable, as the team not only had to worry about the core Ansible code base but also about the modules that shipped with it.

Each namespace potentially has its dev team comprised of core Ansible contributors, community members, and, in the case of some namespaces, large corporations such as Amazon and Microsoft. Thus, trying to coordinate an Ansible release became a challenge, both in terms of logistics and timing. Some of the technology that Ansible supports changes very fast; for example, both Amazon Web Services and Microsoft Azure introduce new features and add functionality to existing services almost weekly. It didn't make sense for Ansible to potentially wait up to six months to provide an update that adds compatibility issues, which is why the Ansible team decided it was time to decouple the release of what is now known as **Ansible Core**, which is the tooling needed to run **Ansible**; this now comprises over 85 name namespace collections, which are made up of well over 1,000 modules and plugins.

## The life cycle of a release

The release cycle begins with introducing a new major version of `ansible-core`, such as `ansible-core 2.11`. Following this, the latest release of `ansible-core` and its two preceding versions, `ansible-base 2.10` and `Ansible 2.9`, are actively maintained. Development then shifts to the **devel** branch, where new features for `ansible-core` are continuously worked on. During this phase, there's a freeze on adding or updating **collections** in the **Ansible community** package.

Subsequently, a release candidate for the **Ansible community** package is introduced. This undergoes testing, and additional release candidates are rolled out if necessary.

Once finalized, a new major version of the **Ansible community** package is released, which aligns with the recent `ansible-core`; for instance, `Ansible 4.0.0` would be based on `ansible-core 2.11`. After this release, only the latest version of the **Ansible community** package remains under active maintenance.

The focus then shifts to **collections**, where new features are developed. Individual **collections** instances have the flexibility to introduce several minor and major versions. On a regular schedule, minor updates are released every four weeks for the three supported `ansible-core` versions, such as `2.11.1`, and for the single supported version of the **Ansible community** package, such as `4.1.0`.

As the cycle progresses, there's a feature freeze on `ansible-core`. This is followed by introducing a release candidate for `ansible-core`, which undergoes testing. If needed, more release candidates are introduced. Finally, the subsequent major version of `ansible-core` is released, marking the commencement of a new cycle.

The following chart provides an overview of the cycle:

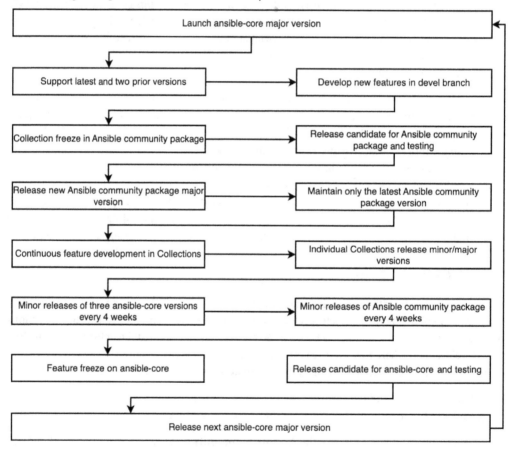

Figure 2.1 – An overview of the Ansible release cycle

As you can see, this approach allows the Ansible team to be a lot more flexible with their release schedules and allows a lot more concurrent work on the two different releases.

Now that we have an idea of how Ansible manages its release cycle and also how modules can be packaged, let's take a look at Ansible Galaxy, which can be used to distribute collections and roles.

## Introduction to Ansible Galaxy

Most people's first exposure to Ansible Galaxy is the website hosted at `https://galaxy.ansible.com/`. The website is home to community-contributed roles and modules:

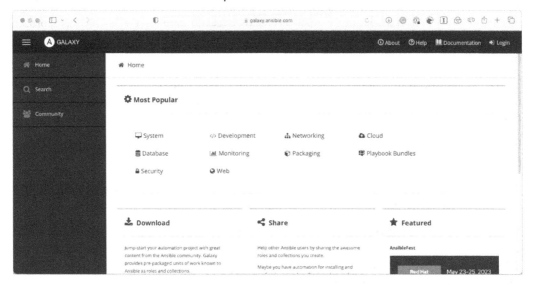

Figure 2.2 – The Ansible Galaxy home page

Throughout the remainder of this book, we will be writing custom roles that interact with the Ansible Core modules for use in our playbook.

More than 15,000 roles are published on Ansible Galaxy; these roles cover many tasks and support almost all the operating systems supported by Ansible.

Then, we have the `ansible-galaxy` command; this is a way of interacting with the Ansible Galaxy website from the comfort of your command line, as well as being able to bootstrap roles, which we will look at shortly; we can also use it to download, search, and publish our custom roles on Ansible Galaxy.

Finally, Red Hat has open-sourced the code for Ansible Galaxy, meaning you can also run a self-hosted version of the site should you need to distribute your roles behind a company firewall.

Before we look at publishing roles and using roles from Ansible Galaxy, let's discuss what a role means.

# What is a role?

Throughout the remainder of this book, we will be building custom roles of our own, so this will be just an overview of what a role is.

In *Chapter 1*, *Installing and Running Ansible*, our final playbook comprised some variables, a handler, four tasks, and a template.

Apart from the template file, all of the code was hardcoded into our playbook file, which, although this makes it easy to read when using a small number of tasks and variables, etc., taking this approach doesn't make the code very re-useable. Additionally, in later chapters, we could potentially be executing over 50 tasks in a single playbook run, which will make for quite a large, unruly file.

To get around this, Ansible has the concept of roles; they allow you to structure your Ansible code in a way that makes sense logically, for example, grouping tasks that perform a single job, which, in *Chapter 1*, *Installing and Running Ansible*, was installing and configuring the NTPD service.

It also means you can drop a role into another playbook by copying the role folder, publishing it, and then pulling it down from Ansible Galaxy.

So, let's look at creating a basic role based on the tasks, handler, variables, and template from the final playbook we ran at the end of *Chapter 1*, *Installing and Running Ansible*.

To start with, we will need to create the folder and file structure recommended by Ansible for a role; luckily, the `ansible-galaxy` command has us covered here; by running the following command in the folder where your playbook is going to be stored, it will bootstrap the folder and file structure, which is considered by Red Hat to be a best practice:

```
$ ansible-galaxy role init roles/learnansible-example-role
```

The preceding command will create a folder called `roles` if one doesn't already exist, and inside the `roles` folder, add a second called `learnansible-example-role`.

The `learnansible-example-role` folder contains all the best-practice folder layouts and files needed to be able to publish a role on Ansible Galaxy.

These are as follows:

- `README.md`: This file contains an outline for you to fill in to provide information on your role; you can use as much or as little as you want of the template. Please note its contents will appear on Ansible Galaxy if you decide to publish your role there, so make it as descriptive as possible.

- `defaults/main.yml`: This YAML file typically contains any default values for your role.

- `files/`: This empty folder holds any files that need to be copied for your role to the target hosts during playbook execution.

- `handlers/main.yml`: As you may have already guessed by the name of this folder, this YAML file is where you define any handlers your role needs.

- `meta/main.yml`: This YAML file, like the `README.md` file, is only used once the role is published to Ansible Galaxy; here, you can provide your details, any tags you want to add, and define the supported platform and the minimum version of Ansible version your role support.

- `tasks/main.yml`: This is the file we will spend most of the time in throughout the rest of the chapters; it is where all of the roles' tasks are defined.

- `templates/`: This is another empty folder; this time, it is here to store your template files.

- `tests/inventory` and `test.yml`: Here, we have a file that contains two files, an inventory file and a test playbook; it is used to run tests on your role.

- `vars/main.yml`: Finally, this YAML file contains any variables that you may wish to use, and these override the contents of the `defaults/main.yml` file, should you need to do so.

To populate the role, I have taken the code from the final playbook and split it across the aforementioned various files; the only change that I made to the playbook itself was to remove the following task, as we don't need it:

```
- name: "Output some information on our host"
  ansible.builtin.debug:
      msg: "I am connecting to {{ ansible_nodename }} which is
running {{ ansible_distribution }} {{ ansible_distribution_version }}"
```

This leaves the `roles/learnansible-example-role/tasks/main.yml` file looking like the following code:

```
# tasks file for roles/learnansible-example-role

- name: "Update all packages to the latest version"
  ansible.builtin.apt:
    name: "*"
    state: "latest"
    update_cache: true
  tags:
    - "skip_ansible_lint"

- name: "Install packages"
  ansible.builtin.apt:
    state: "present"
    pkg:
      - "ntp"
      - "sntp"
      - "ntp-doc"
```

```
- name: "Configure NTP"
  ansible.builtin.template:
    src: "./ntp.conf.j2"
    dest: "/etc/ntp.conf"
    mode: "0644"
  notify: "Restart ntp"
```

Notice that, as mentioned in *Chapter 1, Installing and Running Ansible*, we have - - - at the top of the file to show that main.yml is a separate file. As it is in the tasks folder, we do not need to define that it contains tasks by using tasks, as we did in our original playbook.

This pattern is followed by the roles/learnansible-example-role/handlers/main. yml file, which looks like this:

```
# handlers file for roles/learnansible-example-role

- name: "Restart ntp"
  ansible.builtin.service:
    name: "ntp"
    state: "restarted"
```

Additionally, this is followed by the roles/learnansible-example-role/vars/main. yml file, which contains the following:

```
---
# vars file for roles/learnansible-example-role
ntp_servers:
    - "0.uk.pool.ntp.org"
    - "1.uk.pool.ntp.org"
    - "2.uk.pool.ntp.org"
    - "3.uk.pool.ntp.org"
```

The roles/learnansible-example-role/vars/ntp.conf.j2 file is an exact copy of the template file we used in *Chapter 1, Installing and Running Ansible*.

The only addition outside of the README.md file is roles/learnansible-example-role/ meta/main.yml. This file, as mentioned, contains all of the information needed to publish the role to Ansible Galaxy; in our example, this looks like the following:

```
galaxy_info:
    role_name: "ansible_role_learnansible_example"
    namespace: "russmckendrick"
    author: "Russ McKendrick"
    description: "Example role to accompany Learn Ansible (Second
Edition)"
```

```
    issue_tracker_url: "https://github.com/russmckendrick/ansible-role-
  learnansible-example/issues"
    license: "license (BSD-3-Clause)"
    min_ansible_version: "2.9"
    platforms:
      - name: "Ubuntu"
        versions:
          - "jammy"
    galaxy_tags:
      - "ntp"
      - "time"
      - "example"
  dependencies: []
```

We will revisit this file in the next section of this chapter when it comes to publishing our role to Ansible Galaxy.

Now that we have everything that we need to run the role in place, we need a playbook to call it; in the Chapter02 folder for the repo, which accompanies this title, you will file the roles folder, as described earlier, and a playbook called playbook01.yml, which looks like the following:

```
- name: "Run the role locally"
  hosts: "ansible_hosts"
  gather_facts: true
  become: true
  become_method: "ansible.builtin.sudo"

  roles:
    - learnansible_example_role
```

As you can see, the start of the playbook looks precisely the same as the one we ran at the end of *Chapter 1, Installing and Running Ansible*. However, it is missing the vars, handlers, and tasks sections, and instead, we are just using a roles section, which contains the single role found at roles/learnansible-example-role.

In the Chapter02 folder, you will find all the files needed to launch a local virtual machine using Multipass.

---

**Important note**

When running the following commands, create a copy of the hosts.example file and call it hosts; once copied, update the newly created file with the IP address of the newly launched VM, as we did in *Chapter 1, Installing and Running Ansible*.

To launch the virtual machine, get its IP address and run the playbook using the following commands:

```
$ multipass launch -n ansiblevm --cloud-init cloud-init.yaml
$ multipass info ansiblevm
$ ansible-playbook -i hosts playbook01.yml
```

This should give you something like the following output:

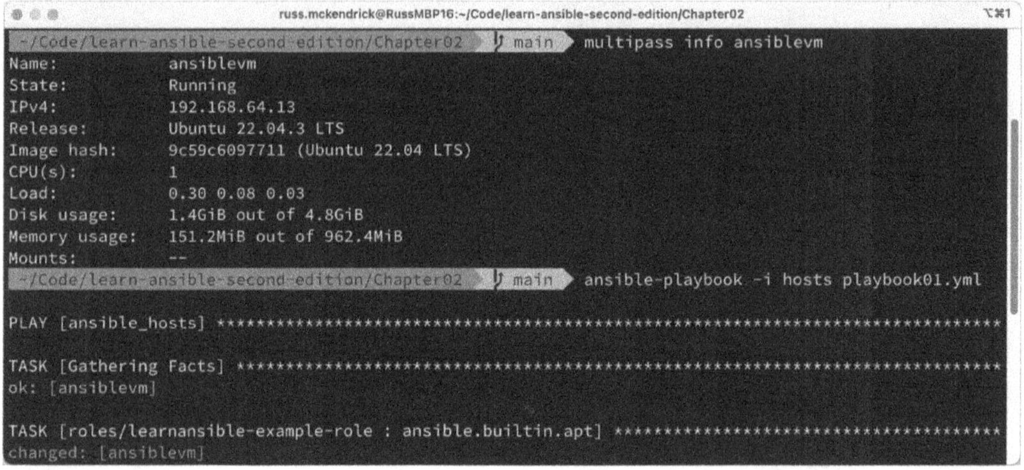

Figure 2.3 – Running the playbook using the newly created role

You can stop and delete the VM by using the following two commands:

```
$ multipass stop ansiblevm
$ multipass delete --purge ansiblevm
```

Now that we know what a role is and have a fundamental one defined, let's look at publishing our simple role to Ansible Galaxy and then use it—along with some others—in an Ansible playbook.

## Publishing to and using Ansible Galaxy roles

Now that we know what a role is and have seen how using them can make our Ansible playbooks a little cleaner and repeatable, we should look at how we can publish our roles to Ansible Galaxy and use them from there in our playbooks.

### Publishing your roles to Ansible Galaxy

You need two main prerequisites when publishing your role to Ansible Galaxy: an active GitHub account, which will be used to authenticate to Ansible Galaxy, and a public GitHub repository containing the code for your role.

In this example, I am going to be using my own GitHub account; you can find me at `http://github.com/russmckendrick/`. I will be using a repository that can be found at `https://github.com/russmckendrick/ansible-role-learnansible-example/`.

To publish your role, you need to take the following steps:

1. Go to the Ansible Galaxy website, which can be found at `https://galaxy.ansible.com/`, and click on the GitHub logo to log in:

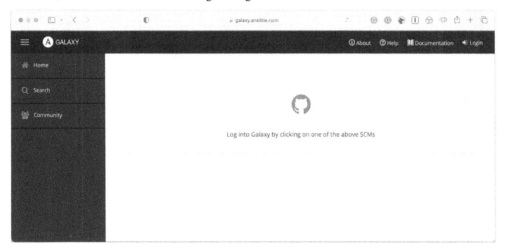

Figure 2.4 – Logging in to Ansible Galaxy

2. Once logged in, click on the **My Content** menu item, which is represented by the bulleted pointed list icon in the left-hand side menu as follows:

Figure 2.5 – Going to the My Content page

3.  Once on the **My Content** page, click on the + **Add Content** button; here, you will be given two options: **Import Role from GitHub** or **Upload New Collection**:

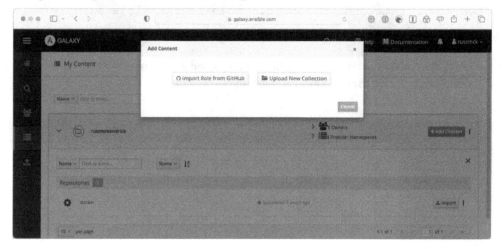

Figure 2.6 – Adding content options

4.  Click on the **Import Role from GitHub** button, and you will be presented with a list of your repositories; select the repository containing the role you would like to publish and click on the **OK** button:

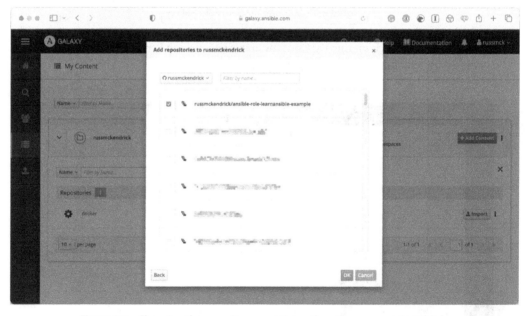

Figure 2.7 – Choosing the repository containing the role you want to publish

5.  After a few moments, your role will be published, and you will be returned to the **My Content** page, which should now list your newly published role:

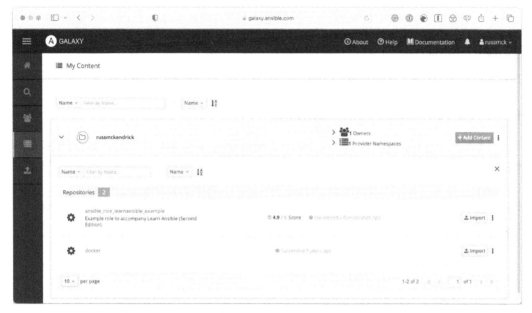

Figure 2.8 – Returning to the My Content page

6.  Click on the newly published role name, which will take you to the Ansible Galaxy roles page:

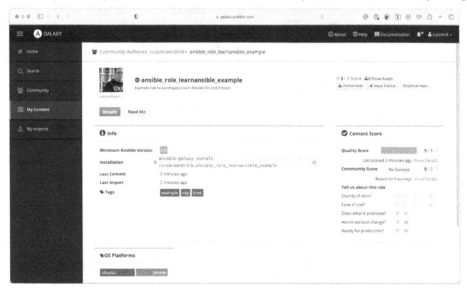

Figure 2.9 – The newly published Ansible Galaxy roles page

You can find a copy of the role I published during this walk-through at `https://galaxy.ansible.com/russmckendrick/ansible_role_learnansible_example`.

As you can see, the details in the **Info** section of the Ansible Galaxy roles page contain the information we defined in the `meta/main.yml` file and clicking on the **Read Me** button on the page will display the rendered contents of the `README.md` file.

Now that the role has been published, how do we use it in our Ansible playbooks? Let's find out.

## Using roles from Ansible Galaxy

The first thing we need to do before we use the role in our playbook is to download the role; there are a few ways of doing this; first, you can download the role using the command given on the roles Ansible Galaxy page.

Running the following command will download the role to your Ansible configuration directory:

```
$ ansible-galaxy install russmckendrick.ansible_role_learnansible_
example
```

The Ansible configuration directory is typically a hidden folder in your user's home folder. The shorthand for this folder is `~/.ansible`, or in my case, the full path to the folder is `/Users/russ.mckendrick/.ansible`, as you can see in the following shell output:

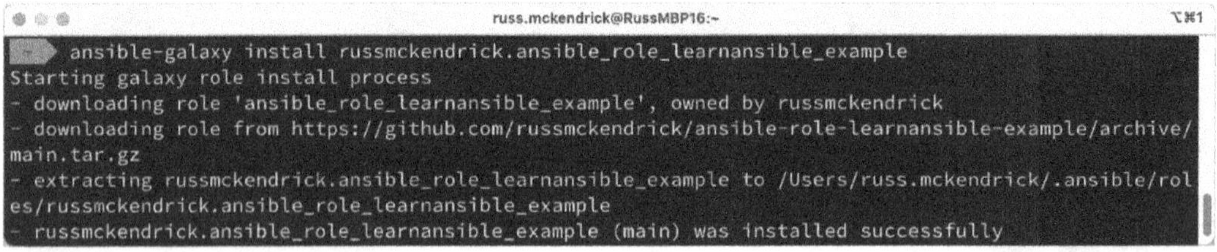

Figure 2.10 – Downloading the role from Ansible Galaxy

The second way to download roles from Ansible Galaxy is to create a `requirements.yml` file; this file should contain a list of the roles you wish to download, for example, the `requirements.yml` file in the `Chapter02` folder of the repository, which accompanies this book and looks like the following:

```
- src: "itnok.update_ubuntu"
- src: "geerlingguy.nginx"
- src: "russmckendrick.ansible_role_learnansible_example"
```

As you can see, there are three roles defined in there; to install all three, you can run the following command:

```
$ ansible-galaxy install -r requirements.yml
```

The two others we will download are the following:

- `itnok.update_ubuntu`: This role manages updates on Ubuntu hosts
- `geerlingguy.nginx`: This role helps you download, install, and configure NGINX on multiple Linux distributions

You can find links for the roles in the further reading section at the end of the chapter.

This will download only the missing roles; when I ran the command, I got the following output:

```
russ.mckendrick@RussMBP16:~/Code/learn-ansible-second-edition/Chapter02
~/Code/learn-ansible-second-edition/Chapter02  main ±  ansible-galaxy install -r requirements.y
ml
Starting galaxy role install process
- downloading role 'update_ubuntu', owned by itnok
- downloading role from https://github.com/itnok/ansible-role-update-ubuntu/archive/2.0.1.tar.gz
- extracting itnok.update_ubuntu to /Users/russ.mckendrick/.ansible/roles/itnok.update_ubuntu
- itnok.update_ubuntu (2.0.1) was installed successfully
- adding dependency: itnok.is_ubuntu
- downloading role 'nginx', owned by geerlingguy
- downloading role from https://github.com/geerlingguy/ansible-role-nginx/archive/3.1.4.tar.gz
- extracting geerlingguy.nginx to /Users/russ.mckendrick/.ansible/roles/geerlingguy.nginx
- geerlingguy.nginx (3.1.4) was installed successfully
[WARNING]: - russmckendrick.ansible_role_learnansible_example (main) is already installed - use
--force to change version to unspecified
- downloading role 'is_ubuntu', owned by itnok
- downloading role from https://github.com/itnok/ansible-role-is-ubuntu/archive/1.1.0.tar.gz
- extracting itnok.is_ubuntu to /Users/russ.mckendrick/.ansible/roles/itnok.is_ubuntu
- itnok.is_ubuntu (1.1.0) was installed successfully
~/Code/learn-ansible-second-edition/Chapter02  main ±
```

Figure 2.11 – Downloading the missing roles from Ansible Galaxy

As you can see, as `russmckendrick.ansible_role_learnansible_example` was already present on my machine, it skipped downloading it.

The Ansible playbook called `playbook02.yml`, which can be found in the `Chapter02` folder, calls the three roles defined in the `requirements.yml` file using the following code:

```
---
- name: "Run the remote roles"
  hosts: "ansible_hosts"
  gather_facts: true
  become: true
  become_method: "ansible.builtin.sudo"

  roles:
    - "itnok.update_ubuntu"
    - "geerlingguy.nginx"
    - "russmckendrick.ansible_role_learnansible_example"
```

As before, you can launch a VM using Multipass (ensuring that you update the IP address in the hosts file) and run the playbook using the following commands:

```
$ multipass launch -n ansiblevm --cloud-init cloud-init.yaml
$ multipass info ansiblevm
$ ansible-playbook -i hosts playbook02.yml
```

As you can see from the playbook recap in the following screen, a lot more happened this time:

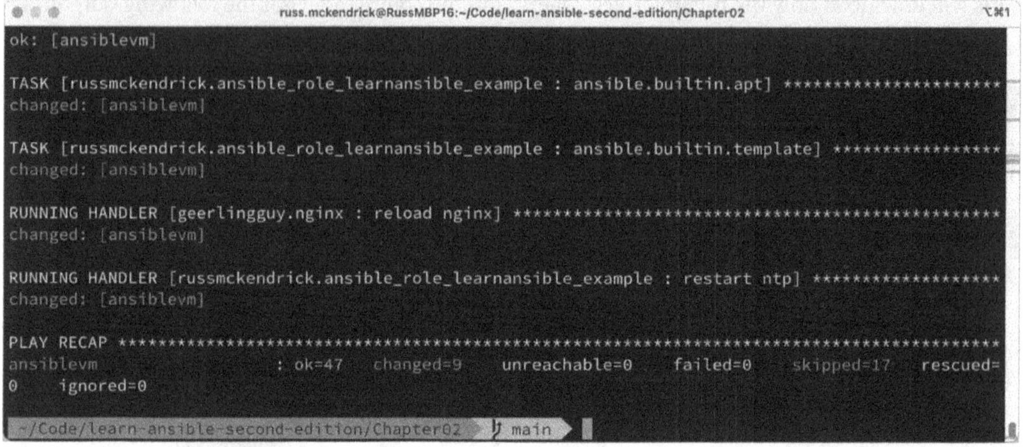

Figure 2.12 – Running the playbook, which uses the roles downloaded from Ansible Galaxy

The two additional roles did a more thorough update of the operating system than our role did, installing the NGINX package and starting the service up; this means that if you put the IP address returned by the `multipass info ansiblevm` command into your browser, you can see the default NGINX page.

Once again, when ready, you can stop and delete the VM by using the following two commands:

```
$ multipass stop ansiblevm
$ multipass delete --purge ansiblevm
```

So now that you understand what an Ansible Role is, how to publish it to Ansible Galaxy, and how to incorporate our own published and community roles into an Ansible playbook, what else can Ansible Galaxy do?

## Ansible collections

At the start of this chapter, we discussed how the Ansible development team decoupled Ansible Modules away from Ansible Core and how this affected the release life cycle.

All of these modules, plugins, and other supporting code are available on Ansible Galaxy; for example, the AWS collection from the Amazon namespace can be found at `https://galaxy.ansible.com/amazon/aws`; you can install the collection using the following command:

```
$ ansible-galaxy collection install amazon.aws
```

Running this command will download and install the collection to `~/.ansible/collections/ansible_collections/amazon/aws`, as seen in the following terminal output:

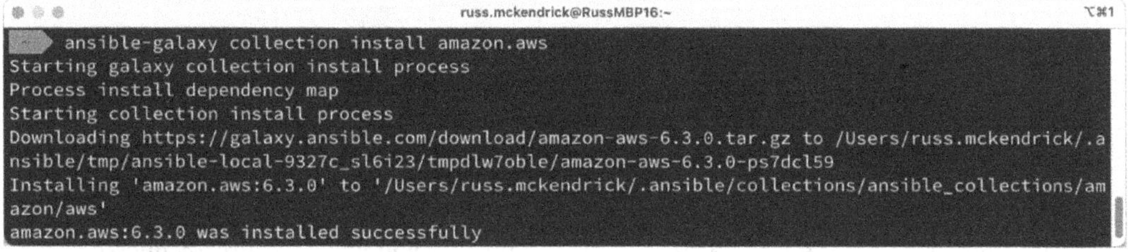

Figure 2.13 – Installing the amazon.aws collection

However, just having the collection installed doesn't mean that you will be able to use it within your playbooks just yet; for example, the Amazon AWS module requires some additional Python libraries to be installed, and typically, each collection will come with a `requirements.txt` file that lists the required Python libraries that need to be installed on your system for the collections modules and plugins to work.

To install these libraries, you should use pip to install them:

```
$ pip install -r ~/.ansible/collections/ansible_collections/amazon/
aws/requirements.txt
```

Once installed, you will be able to use the modules and plugins which make up the collection.

## Ansible Galaxy commands

Before finishing this chapter on Ansible Galaxy, let's quickly discuss some other useful commands.

The `ansible-galaxy` command has some of the basic functionality you would expect, such as the following:

```
$ ansible-galaxy --version
$ ansible-galaxy --help
```

This displays details of the version and basic help options on the command, respectively.

The `ansible-galaxy` command is split into two parts, which we have already touched upon.

First, there is `ansible-galaxy collection`; from here, you can add the following commands:

- `download`: Retrieves collections and their dependencies, such as **tarballs**, which is an archive format for Linux machines for offline installations.
- `init`: Set up a new collection with the foundational structure.
- `build`: Construct an Ansible collection artifact suitable for publication to Ansible Galaxy.
- `publish`: Release a collection artifact to Ansible Galaxy.
- `install`: Add collection(s) from the specified file(s), URL(s), or directly from Ansible Galaxy.
- `list`: Display the name and version of every collection in the collections path.
- `verify`: Contrast the checksums of the installed collection(s) with those on the server; any dependencies are not verified.

Secondly, there is `ansible-galaxy role`, as I am sure you will have already guessed; these commands are for working with roles:

- `init`: Set up a new role with the foundational structure of a role
- `remove`: Erase roles from the specified roles path
- `delete`: Remove the role from Galaxy. Note that this does not affect or modify the actual GitHub repository
- `list`: Display the name and version of every role in the roles path
- `search`: Query the Galaxy database using tags, platforms, authors, and keywords
- `import`: Incorporate a role into a Galaxy server
- `setup`: Oversee the connection between Galaxy and the designated source
- `info`: Obtain detailed information about a particular role
- `install`: Add role(s) from specified file(s), URL(s), or directly from Ansible Galaxy

For more help on any of these commands, you can append `--help` to the end. For example, you can run the following:

```
$ ansible-galaxy role search --help
```

This will give you detailed information on how to search Ansible Galaxy; for example, to search for my roles, you would need to run the following:

```
$ ansible-galaxy role search --author russmckendrick
```

This returns a list of all of the roles I have published to Ansible Galaxy, or you could run this:

```
$ ansible-galaxy role list
```

This lists all the roles you have installed in the roles path (which is `~/.ansible/roles/`). From there, you could run something like the following:

```
$ ansible-galaxy role info russmckendrick.ansible_role_learnansible_
example
```

This helps you obtain information on a role you have installed, which concludes our look at the `ansible-galaxy` command.

## Summary

In this chapter, we have had an in-depth look at Ansible Galaxy, the website, and the command-line tool. We have also discussed the Ansible development and release cycle and understood what an Ansible Role is.

I am sure that you will agree that Ansible Galaxy offers valuable community services in that it allows users to share roles for everyday tasks and provides a way for users to contribute to the Ansible community by publishing their roles.

However, just be careful. Remember to check the code and read through bug trackers before using roles from Ansible Galaxy in production environments; after all, many roles need escalated privileges to execute their tasks successfully.

As mentioned in this chapter, we will be creating our own Ansible Roles throughout the remainder of this title, and there will be additional hints and tips on creating and using roles as our Ansible playbooks get more and more sophisticated.

In our next chapter, we will look at more Ansible commands and tools that ship as part of Ansible Core.

## Further reading

You can find more details on the two additional roles that we installed from Ansible Galaxy and the official documentation at the following sites:

- geerlingguy.nginx: `https://galaxy.ansible.com/ui/standalone/roles/geerlingguy/nginx`

- itnok.update_ubuntu: `https://galaxy.ansible.com/ui/standalone/roles/itnok/update_ubuntu/`

- **Ansible Galaxy Documentation**: `https://ansible.readthedocs.io/projects/galaxy-ng/en/latest/community/userguide/`

# 3

# The Ansible Commands

Before moving on to writing and executing more advanced playbooks, we will look at the rest of the built-in Ansible commands. Here, we will cover using the commands that make up Ansible. Toward the end of this chapter, we will install a third-party tool to visualize our host inventory.

This chapter will cover the following topics:

- Inbuilt commands
- Third-party commands

## Inbuilt commands

When we installed Ansible, several different commands were installed. These were as follows:

- `ansible`
- `ansible-config`
- `ansible-console`
- `ansible-doc`
- `ansible-galaxy`
- `ansible-inventory`
- `ansible-playbook`
- `ansible-pull`
- `ansible-vault`

We already covered the `ansible-galaxy` command in *Chapter 2, Exploring Ansible Galaxy*. We will be looking at `ansible-playbook` throughout the remaining chapters of this book, so I will not go into any detail about that command in this chapter. Let's start at the top of the list and a command we have already used.

## Ansible

Now, you would have thought that `ansible` would be the most common command we will use throughout this book, but it isn't.

The `ansible` command is only ever used for executing ad hoc commands against a single host or collection of hosts. In *Chapter 1, Installing and Running Ansible*, we created a host inventory file that targeted a single local virtual machine.

For this part of the chapter, we'll look at targeting four different hosts I have running in a cloud provider; my host's file looks as follows:

```
ansible01 ansible_host=139.162.233.174
ansible02 ansible_host=139.162.233.227
ansible03 ansible_host=139.144.132.49
ansible04 ansible_host=139.144.132.71
[london]
ansible01
ansible02
[nyc]
ansible03
ansible04
[demohosts:children]
london
nyc
[demohosts:vars]
ansible_connection=ssh
ansible_user=root
ansible_private_key_file=~/.ssh/id_rsa
host_key_checking=False
```

As you can see, I have four hosts – `ansible01` > `ansible04`. My first two hosts are in a group called `london` and my second two are in a group called `nyc`. I have then taken these two groups and created one containing them called `demohosts`, and I used this group to apply some basic configurations based on the hosts I have launched.

Using the `ping` module, I can check connectivity to the hosts by running the following commands. First, let's check the two hosts in `london`:

```
$ ansible -I hosts london -m ping
```

This returns the following results:

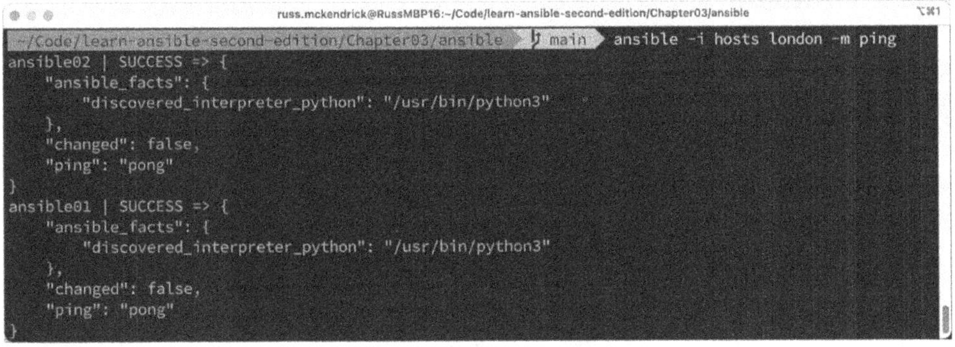

Figure 3.1 – Doing an Ansible ping targeting the london hosts

Now, let's run the same command, but this time targeting the `nyc` hosts:

```
$ ansible -i hosts nyc -m ping
```

This gives us the following output:

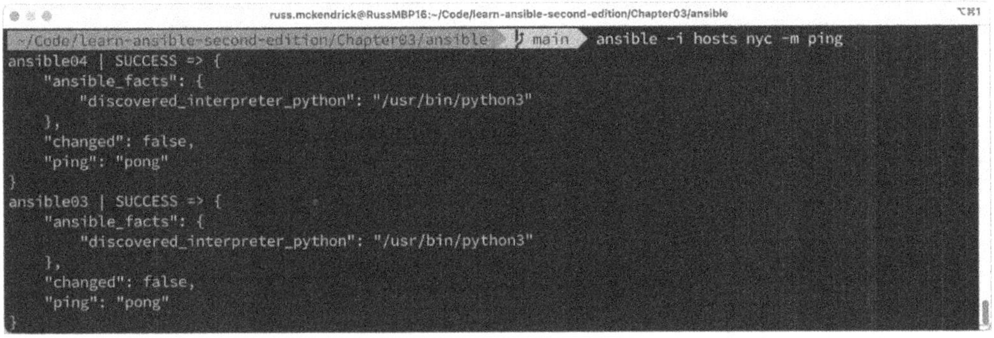

Figure 3.2 – Doing an Ansible ping targeting the nyc hosts

As you can see, all four of my hosts returned `pong`.

I can also target all four hosts at once by adding `all` rather than a particular group of hosts:

```
$ ansible -i hosts all -m ping
```

Now that we can access our host through Ansible, we can target them and run some ad hoc commands; let's start with something basic:

```
$ ansible -i hosts london -a "ping -c 3 google.com"
```

This command will connect to the `london` hosts and run the `ping -c 3 google.com` command; this will `ping` the `google.com` domain from the hosts and return the results:

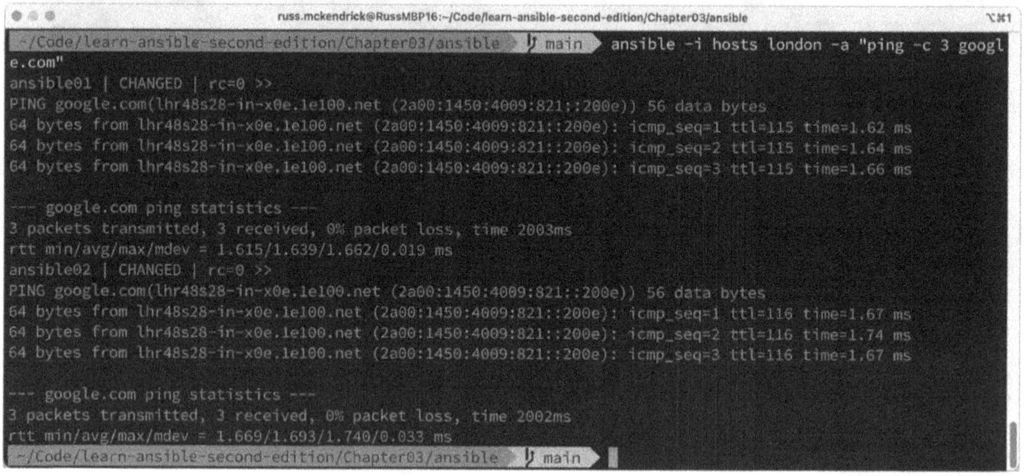

Figure 3.3 – Running the ping command against google.com

We can also run a single module using the `ansible` command; we did this in *Chapter 1, Installing and Running Ansible*, using the `setup` module. However, a better example would be updating all the installed packages across all the hosts by running the following command:

```
$ ansible -i hosts all -m ansible.builtin.apt -a "name=* state=latest
update_cache=ye"
```

As you can see, we have taken the `ansible.builtin.apt` module, which we defined as follows in *Chapter 1, Installing and Running Ansible*:

```
- ansible.builtin.apt:
    name:"*"
    state:"latest"
    update_cache:"true"
```

I've passed in the same options, but rather than use YAML, I have formatted it as a key and value, which is typical of what you would pass into any command on the command line:

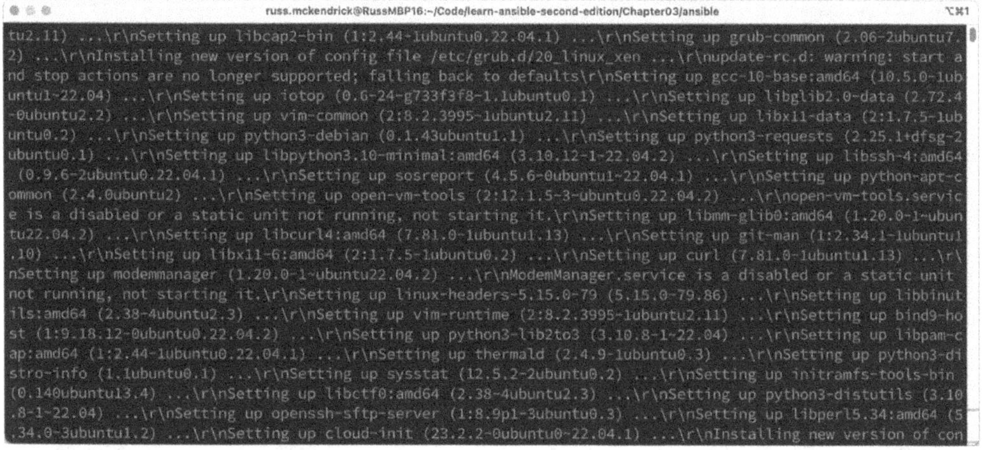

Figure 3.4 – Using the ansible.builtin.apt module to update all the packages

As you can see, the output when running Ansible is quite verbose, and it provides feedback to tell us precisely what it did during the ad hoc execution.

Let's rerun the command against all our hosts, but this time just for a single package, say `ntp`:

```
$ ansible -i hosts all -m ansible.builtin.apt -a "pkg=ntp
state=latest"
```

Running the command once will install the package on all four of our hosts:

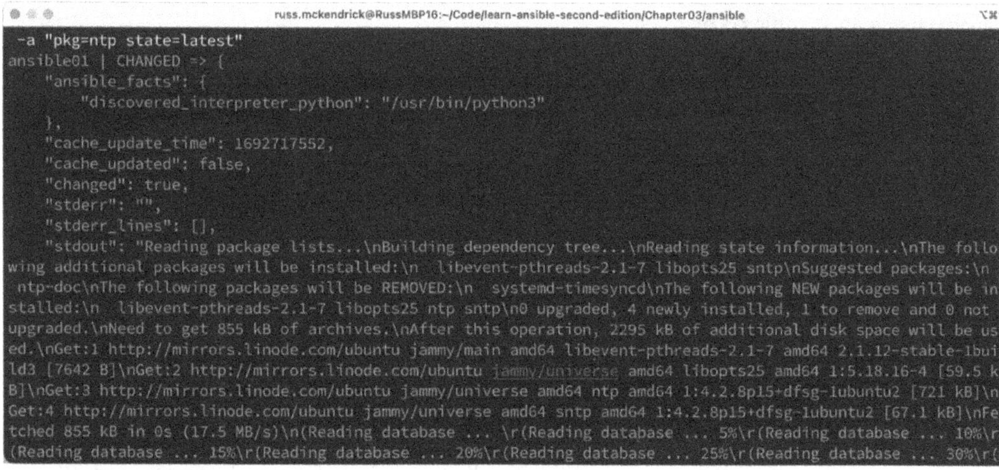

Figure 3.5 – Using the ansible.builtin.apt module to install the ntp package

Now, let's rerun the command:

```
$ ansible -i hosts all -m ansible.builtin.apt -a "pkg=ntp
state=latest"
```

Running the command once will install the package on all four of our hosts and give us the following results:

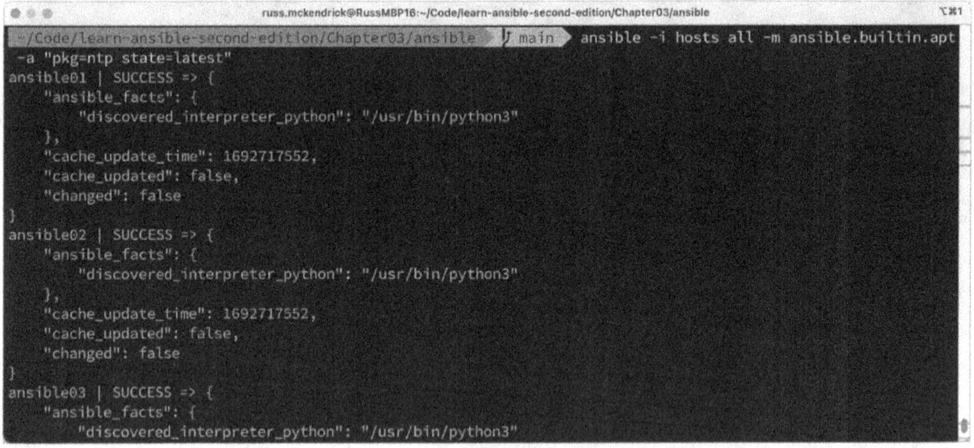

Figure 3.6 – Rerunning the ansible.builtin.apt module to install the ntp package

As you can see, the hosts are returning a SUCCESS status and are showing no changes, which is what we would expect to see.

So, why would you want to do this, and what is the difference between the two commands we ran?

First, let's take a look at two of the commands we initially ran once we confirmed our hosts were available using an Ansible ping:

```
$ ansible -i hosts london -a "ping -c 3 google.com"
$ ansible -i hosts all -m ansible.builtin.apt -a "name=* state=latest
update_cache=true"
```

While it appears that the first command isn't running a module, it is. The default module for the ansible command is called raw and runs raw commands on each of the targeted hosts. The -a part of the command passes arguments to the module. The raw module happens to accept raw commands, which is precisely what we are doing with the second command.

As mentioned previously, you will have noticed that the syntax is slightly different when we pass commands to the ansible command and when using it as part of a YAML playbook. All we are doing here is passing the key-value pairs directly to the module.

So, why would you want to use Ansible like this? Well, it's excellent for running commands directly against non-Ansible managed hosts in an extremely controlled way.

Ansible uses SSH to connect to the hosts, runs the command, and lets you know the results. Just be careful – it is easy to get overconfident and run something like the following:

```
$ ansible -I hosts all -a "reboot now"
```

If the user Ansible is using to connect to the host has permission to execute the command, it will just run the command you give it. Running the previous command will reboot all the servers in the host inventory file:

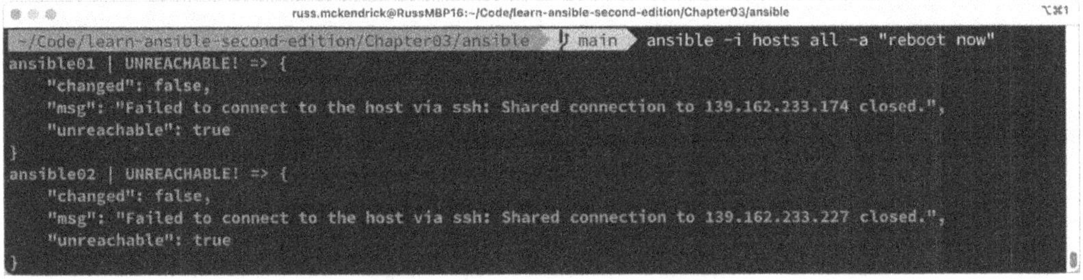

Figure 3.7 – Rebooting all four of the hosts with a single command

All hosts have an UNREACHABLE status because the reboot command kicked our SSH session before the SUCCESS status could be returned. You can, however, see that each of the hosts has been rebooted by running the uptime command:

```
$ ansible -i hosts all -a "uptime"
```

The following screenshot shows the output for the preceding command:

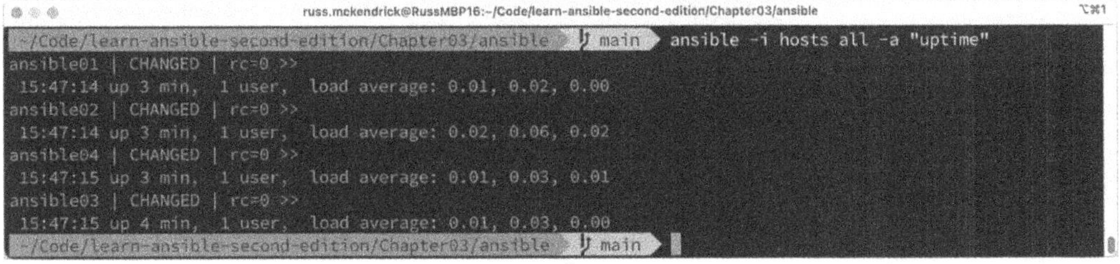

Figure 3.8 – Checking the uptime of the four hosts

> **Important**
>
> As mentioned previously, plus speaking from experience (it's a long story), please be extremely careful when using Ansible to manage hosts using ad hoc commands – it's a powerful but dumb tool, and it will assume you know the consequences of running the commands against your hosts.

That concludes our look at the `ansible` command; let's move on to our next command, `ansible-config`.

## The ansible-config command

The `ansible-config` command is used to manage Ansible configuration files. Ansible ships with sensible defaults, so there is little to configure outside of these. You can view the current configuration by running the following:

```
$ ansible-config dump
```

As shown from the following output, all the text in green is the default config, and any configuration in orange is a changed value:

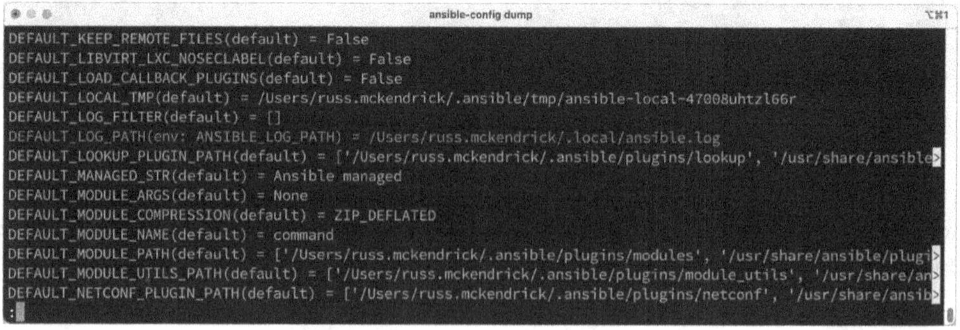

Figure 3.9 – Dumping our complete Ansible configuration to screen

Running the following command will list details of every configuration option there is within Ansible, including what the option does, its current state, when it was introduced, the type, and much more:

```
$ ansible-config list
```

The following screenshot shows the output for the preceding command:

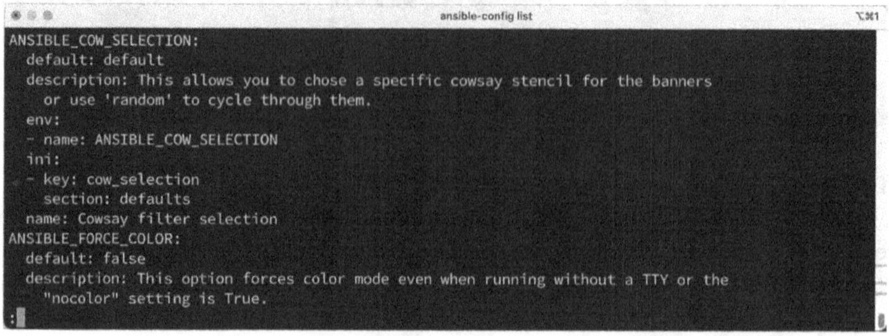

Figure 3.10 – Viewing details on an Ansible configuration option

If you had a configuration file, say at ~/.ansible.cfg, then you can load it using the -c or--config flags:

```
$ ansible-config view --config "~/.ansible.cfg
```

The previous command will give you an overview of the custom configuration file and display the Ansible default values not defined in your custom configuration file.

## The ansible-console command

Ansible has a built-in console. It is not something I have used much in my day-to-day running of Ansible. To start the console, we need to run one of the following commands:

```
$ ansible-console -i hosts
$ ansible-console -i hosts london
$ ansible-console -i hosts nyc
```

The first of the three commands targets all of the hosts, while the next two just target the named groups:

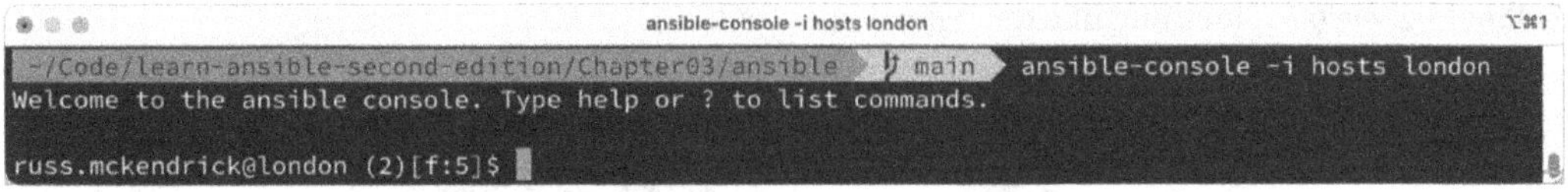

Figure 3.11 – Establishing the console connection

Once connected, you will see that I am connected to the london group of hosts, in which there are two hosts. From here, you can type a module name, such as ping:

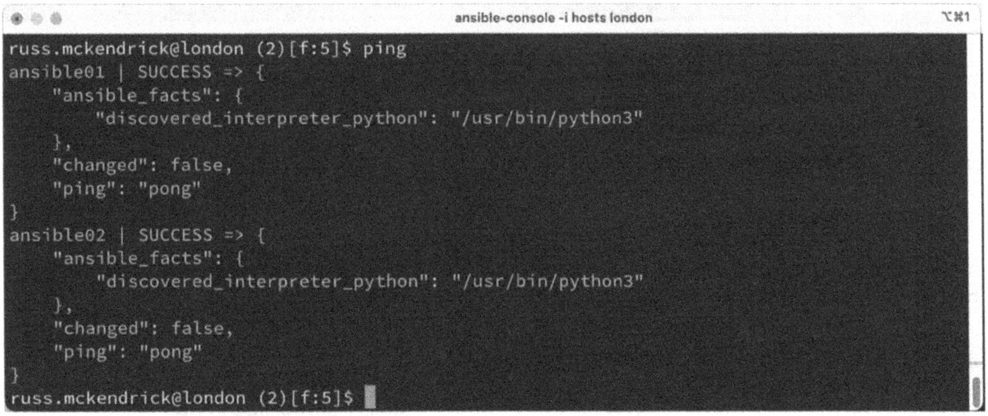

Figure 3.12 – Running ping from Ansible

Alternatively, you can use the `raw` module; for example, you can check the `uptime` command by typing `ansible.builtin.raw uptime`:

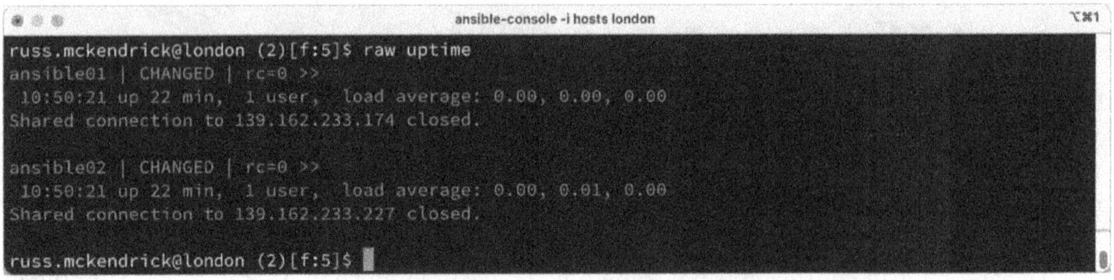

Figure 3.13 – Using the raw module to run the uptime command

You can also use the same syntax as we did when running the `ansible` command to pass key-value pairs – for example, running the following at the console prompt:

```
ansible.builtin.apt pkg=ntp state=latest update_cache=true
```

It should give you something like the following output:

Figure 3.14 – Checking that the ntp package is installed using the ansible.builtin.apt module

You may have noticed that the syntax of the command we are running this time is slightly different from when we ran the same module using the `ansible` command earlier in this chapter.

That command was as follows:

```
$ ansible -i hosts london -m ansible.builtin.apt -a"pkg=ntp
state=latest update_cache=true"
```

Whereas this time, we just ran the following:

```
ansible.builtin.apt pkg=ntp state=latest update_cache=true
```

The reason for this is that when we called the module using the `ansible` command, we were working on the command line of our local machine, so we needed to pass in the module name using the

-m flag and then define the attributes by using the -a flag. After, we had to pass in our key-value pairs within quotation marks so as not to break the flow of the command as spaces are used as a delimiter when it comes to the command line.

When we ran the Ansible console, we had effectively already run the `ansible -i hosts london` part of the command, left our local command line altogether, and were interacting with Ansible itself directly.

To leave the console, type `exit` to return to your regular command-line shell.

As mentioned at the start of this section, the `ansible-console` command is something I do not use – mainly for the warning I gave when we looked at the `ansible` command at the start of this chapter.

When connecting to several hosts using the `ansible-console` command, you must be 100% confident that the commands you are typing are correct. For example, while I was only connected to two hosts, my `hosts` file could have contained 200 hosts. Now, imagine I typed the wrong command – executing it across 200 hosts at once could potentially do some unwanted things, such as rebooting them all simultaneously.

To quit the `ansible-console` session, simply type `exit` and hit *Enter*.

As you have probably guessed, this happened to me. It wasn't 200 hosts, but it could have easily been – *so please be careful.*

## The ansible-inventory command

Using the `ansible-inventory` command provides you with details of your host inventory files. It can be helpful to understand how your hosts are grouped. For example, let's say I run the following command:

```
$ ansible-inventory -i hosts--graph
```

In the same folder as the `hosts` inventory file that I have been using throughout this section, the following is returned:

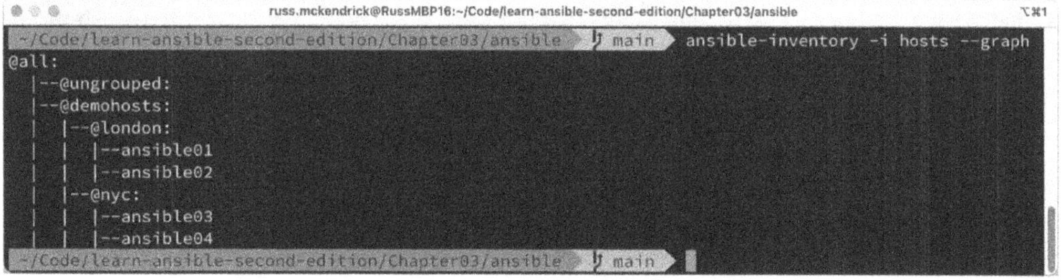

Figure 3.15 – Getting an overview of the inventory hosts file

As you can see, it displays the groups, starting with `all`, then the main host group (`demohosts`), then the child groups (`london` and `nyc`), and finally the hosts themselves (`ansible01` > `ansible04`).

If you want to view the configuration for a single host, you can use this command:

```
$ ansible-inventory -i hosts --host=ansible01
```

The following screenshot shows the output of the preceding command:

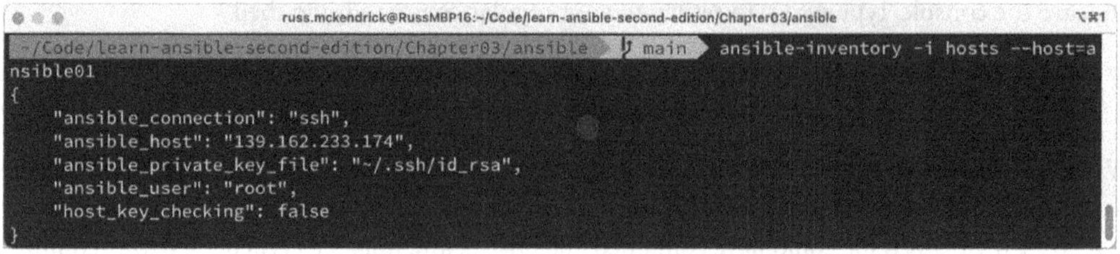

Figure 3.16 – Viewing a single host

You may have noticed that it displays the configuration information that the host inherited from the configuration we set for all the hosts in the `demohost` host group in the inventory file. You can view all the information on each of your hosts and groups by running the following command:

```
$ ansible-inventory -i hosts --list
```

This command is helpful if you have a large or complicated host inventory file and want information on just a single host or if you have taken on a host inventory and want a better idea of how the inventory is structured. We will look at a third-party tool later in this chapter that gives more display options.

## What is ansible-pull?

Like the `ansible-console` command, `ansible-pull` is not a command I use very often; I can count on one hand how often I have used it in the past several years.

`ansible-pull` is a command that allows a target machine to pull its configuration from a given source, such as a Git repository, and apply it locally. This reverses the typical Ansible push model, where a central control node pushes configuration to managed nodes.

The `ansible-pull` command works as follows:

1. The target machine, the one running `ansible-pull`, fetches a specified repository.
2. Once the repository has been fetched, the target machine looks for a playbook. By default, it looks for one called `localhost.yml`, but you can specify a different playbook file if you need to – please note that this is not included in the example files.
3. The target machine then runs the playbook against itself.

There are a few use cases for `ansible-pull`:

- **Decentralized configuration management**: In environments where a centralized Ansible server might be a single point of failure or isn't feasible, `ansible-pull` allows nodes to self-configure by pulling their configurations

- **Edge locations and remote sites**: For edge locations or remote sites with limited connectivity, `ansible-pull` can be scheduled to run at specific intervals via a cron job, ensuring that hosts can self-update when they have connectivity

- **Development and testing**: Developers can use `ansible-pull` to pull down and apply configurations to their local development environments, ensuring consistency with production configurations

There are a few prerequisites to running `ansible-pull` – the most prominent being that the host running `ansible-pull` must have an active and valid Ansible installation and any other dependencies needed to execute the playbook.

In summary, `ansible-pull` provides a way to invert the traditional Ansible model, allowing hosts to pull their configurations as needed rather than having a central host push the configurations to them, as we did in *Chapter 1, Installing and Running Ansible*, and *Chapter 2, Exploring Ansible Galaxy*. For the remainder of this book, we will be taking the more traditional approach to Ansible deployments and pushing our configuration to our target hosts.

However, it is always good to know that if, for whatever reason, you are not able to take this approach, then you do have an alternative option in `ansible-pull`.

## Using the ansible-vault command

In Ansible, it is possible to load variables from files or within a playbook itself; we will look at this in the next chapter in more detail. These files can contain sensitive information such as passwords and API keys. Here's an example:

```
secret:"mypassword"
secret-api-key:"myprivateapikey"
```

As you can see, we have two sensitive bits of information visible as plaintext. This is OK while the file is on our local machine – well, just about OK. But what if we want to check the file into source control to share it with our colleagues?

*We shouldn't store this information in plaintext, even if the repository is private.*

Ansible introduced Ansible Vault to help solve this very problem. Using the `ansible-vault` command, we can encrypt a file or just variables, and then when Ansible is executed, it can be decrypted in memory, and the content can be read as part of the execution.

> **Note**
>
> For the rest of the chapter, I will set a Vault password of `password`, should you wish to run the `ansible-vault` command against the files in the `Chapter03/vault` folder.

To encrypt a file, we need to run the following command, providing a password that will be used to decrypt the file when prompted:

```
$ ansible-vault encrypt secrets.yml
```

The following screenshot shows the output of the preceding command:

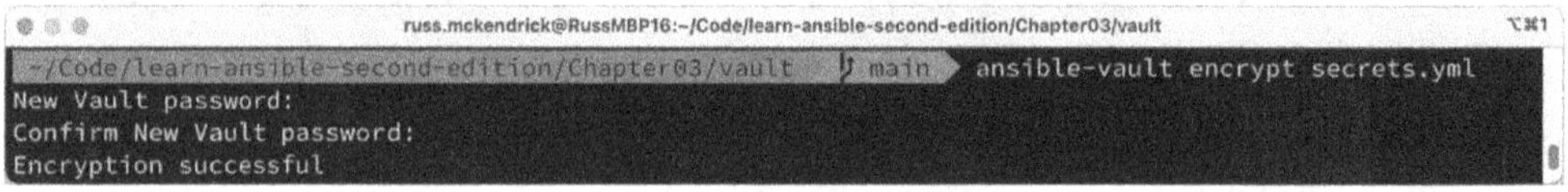

Figure 3.17 – Using ansible-vault to encrypt an entire file

As you can see from the output, you will be asked to confirm the password. Once encrypted, your file will look something like this:

```
$ANSIBLE_VAULT;1.1;AES256
623731386430386366643631666466373331313864313661376436303264333032313
36331303262
366138306131343665346466303962633876233646630310a3064376664623134396
36634646633
396534353334333263613065313938326130383536653338663831613132393431343
76632316263
373663366530316163a39326563306666363133623961393836313303062626136333
33030336430
663438333765323138663638386534643830656337376137353237393032323830313
26262376366
616631366234313063636663303738313362303231323362636262373665393261623
73564353937
303465306233636336333303533633232623233
```

As you can see, the details are encoded using text. This ensures that our `secrets.yml` file will still work without problems when it's checked into source control such as Git.

You can view the content of a file by running the following command:

```
$ ansible-vault view secrets.yml
```

This will ask you for the password and print the content of the file to the screen:

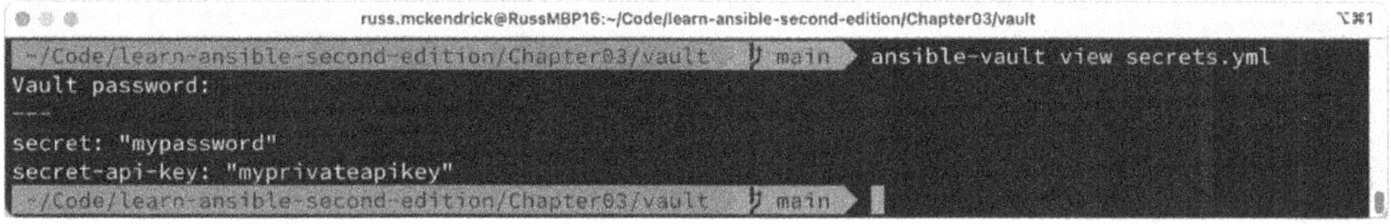

Figure 3.18 – Using ansible-vault to encrypt an entire file

You can decrypt the file on disk by running the following command:

```
$ ansible-vault decrypt secrets.yml
```

This will restore the file to its unencrypted original state.

> **Important**
>
> When using the `ansible-vault decrypt` command, please do not commit or check the decrypted file into your source control system!

Since early in the release of Ansible 2, encrypting a single variable in a file is now possible. Let's add some more variables to our file:

```
username: "russmckendrick"
password: "mypassword"
secretapikey: "myprivateapikey"
packages:
  - apache2
  - ntp
  - git
```

It would be good if we didn't have to keep viewing or decrypting our file to check its variable name and overall content.

Let's encrypt the password content by running the following command:

```
$ ansible-vault encrypt_string'mypassword'--name'password'
```

This will encrypt the `mypassword` string and give it a variable name of `password`:

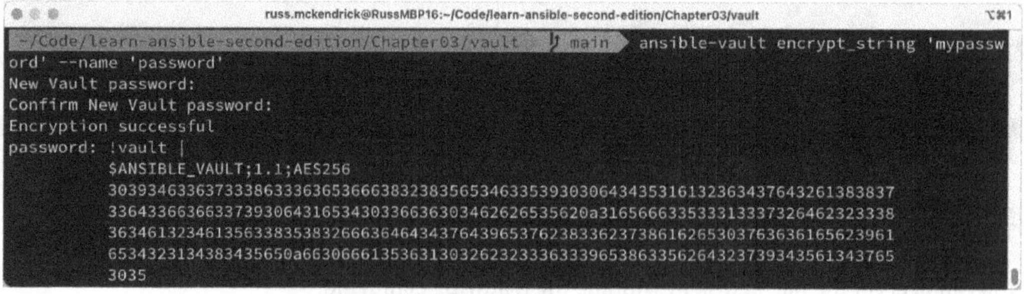

Figure 3.19 – Using ansible-vault to encrypt a single string

We can then copy and paste the output into our file and repeat this process for `secretapikey`:

```
$ ansible-vault encrypt_string 'myprivateapikey' --name 'secretapikey'
```

With that, we have generated two secret variables and replaced the unencrypted ones in our variables file.

> **Note**
>
> For ease of reading, I have truncated the output a little – the entire file can be found in the `Chapter03/vault` folder in this book's GitHub repository.

Our variables file should end up looking something like this:

```
username:"russmckendrick"
password: !vault |
          $ANSIBLE_VAULT;1.1;AES256
30393463363733386333363653666382383565346335393030643435316132363437643
43261383837
          3035
secretapikey: !vault |
          $ANSIBLE_VAULT;1.1;AES256
38663133393834646638663632353634343638626237333438336131316538623737616
66539326263
          3934
packages:
  - apache2
  - ntp
  - git
```

As you can see, that is much easier to read and is just as secure as encrypting the file.

So far, so good, but how do you use Ansible Vault encrypted data in an Ansible playbook?

Before we look at how to do this, let's see what happens when you don't tell the `ansible-playbook` command you are using Ansible Vault by running the following playbook. As you can see, it is loading in the `myvars.yml` file and then printing the contents of our variables to the screen using the `ansible.builtin.debug` module:

```
---


- name: "Print some secrets"
  hosts: "localhost"
  vars_files:
    - "myvars.yml"
  tasks:
    - name: "Print the vault content"
      ansible.builtin.debug:
        msg:
          - "The username is {{ username }} and password is {{
password }}, also the API key is {{ secretapikey }}"
          - "I am going to install {{ packages }}"
```

We can run the playbook using the following command; note that since it's just running locally, we are not passing an inventory file. This is something it will give you a warning about:

```
$ ansible-playbook playbook01.yml
```

This results in an error message being shown in the Terminal output:

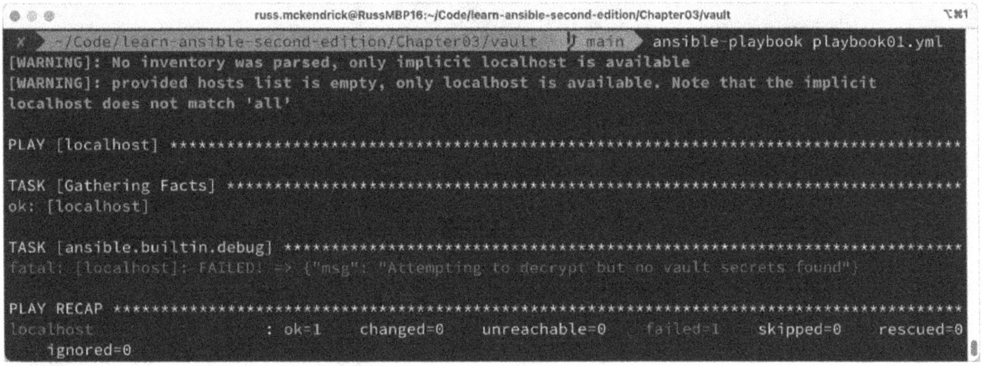

Figure 3.20 – Getting an error when running the ansible-playbook command

As you can see, it's complaining that it found Vault-encrypted data in one of the files, but we haven't provided the secret to unlock it.

The first way we can pass the Vault password during the `ansible-playbook` run is to put the password in a text file and have the `ansible-playbook` command read the file's contents.

As mentioned at the start of this section, I have been encoding my Vaults using a password of `password`. Let's put that in a file and then use it to unlock our Vault:

```
$ echo "password" > /tmp/vault-file
```

Running the following command will read the content of `/tmp/vault-file` and decrypt the data:

```
$ ansible-playbook --vault-id /tmp/vault-file playbook01.yml
```

As you can see from the following playbook run, the output is now as we expect:

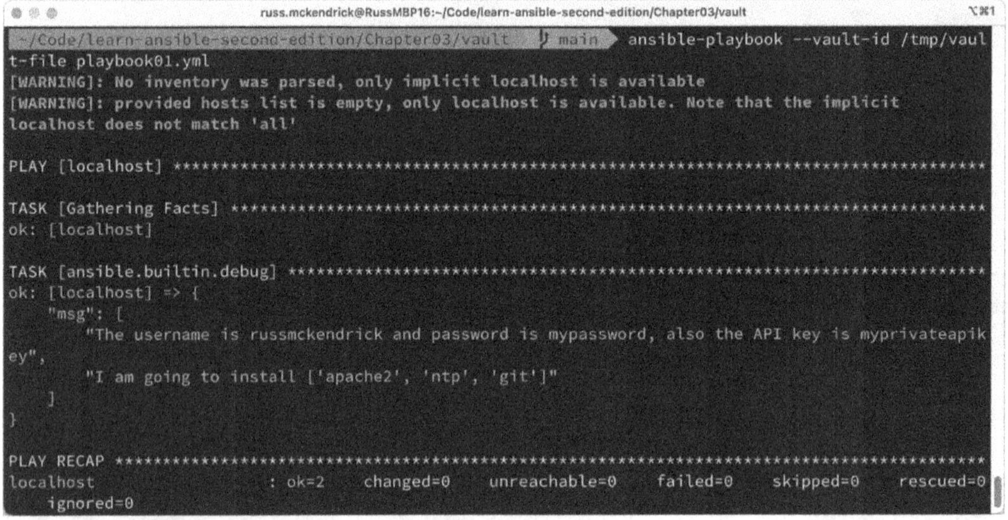

Figure 3.21 – Running ansible-playbook and passing the Vault password in via a file

If you prefer to be prompted for the password, you can use the following command:

```
$ ansible-playbook --vault-id @prompt playbook01.yml
```

The following output shows the prompt:

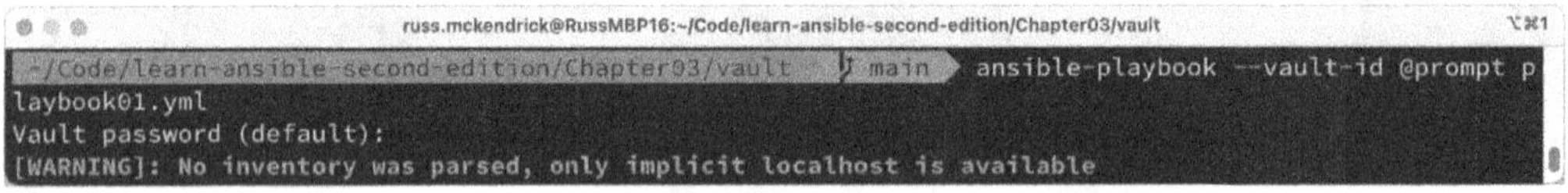

Figure 3.22 – Running ansible-playbook and entering the password via a prompt

You might be asking yourself, why are there two different options? When prompted, just running the command and entering the password might seem enough.

However, when it comes to using services such as the ones we will cover in *Chapter 15, Using Ansible with GitHub Actions and Azure DevOps*, the commands need to run utterly unattended as there will not be an active terminal for you to enter the password.

Another advantage that will be looked at in both *Chapter 15, Using Ansible with GitHub Actions and Azure DevOps*, and *Chapter 16, Introducing Ansible AWX and Red Hat Ansible Automation Platform*, is that by abstracting away the need for an end user to enter credentials at runtime, it is entirely possible for someone to run a pipeline and never need to know or have access to any of the secrets stored in your playbooks or the credentials to unlock them.

# Third-party commands

Before we finish looking at the various Ansible commands, let's look at a command that isn't shipped as part of Ansible itself but is, in fact, a third-party open source project.

## The ansible-inventory-grapher command

The `ansible-inventory-grapher` command, by Will Thames, uses the Graphviz library to visualize your host inventories. The first thing we need to do is install Graphviz. To install this on macOS using Homebrew, run the following command:

```
$ brew install graphviz
```

To install Graphviz on Ubuntu, use the following command:

```
$ sudo apt-get install graphviz
```

Once installed, you can install `ansible-inventory-grapher` using `pip`:

```
$ pip install ansible-inventory-grapher
```

Now that we have everything installed, we can generate the graph using the `hosts` file we used earlier in this chapter:

```
$ ansible-inventory-grapher -i hosts demohosts
```

This will generate something that looks like this:

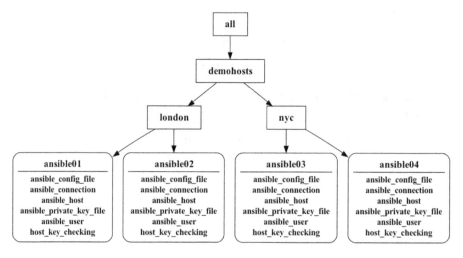

Figure 3.23 – Running ansible-inventory-grapher against our hosts file

This is the raw output of the graph. As you can see, it is like and uses some of the same syntax as HTML. We can render this using the `dot` command, which ships as part of Graphviz. The `dot` command creates hierarchical drawings from graphs. To do this, run the following command:

```
$ ansible-inventory-grapher -i hosts demohosts | dot -Tpng > hosts.png
```

This will generate a PNG file called `hosts.png` that contains the visualization of the host inventory file you can see here:

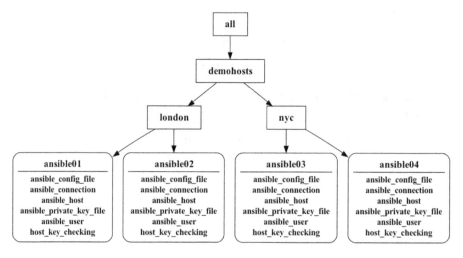

Figure 3.24 – The output of passing our ansible-inventory-grapher output through Graphviz

As you can see, this is an excellent representation of the hosts being targeted by Ansible; it works great for inclusion in your documentation but also gives you an idea of how a complicated inventory file is structured.

## Summary

In this chapter, we briefly looked at some of the supporting tools that ship as part of a standard Ansible installation and a useful third-party tool designed to work with Ansible.

We will use these commands and the one we have purposely missed, `ansible-playbook`, in later chapters.

In the next chapter, we will write a more complex playbook that installs a basic LAMP stack on our local virtual machine.

## Further reading

You can find the documentation for each of the tools covered in this chapter at the following URLs:

- Ansible command-line tools overview: `https://docs.ansible.com/ansible/latest/command_guide/command_line_tools.html`

- ansible: `https://docs.ansible.com/ansible/latest/cli/ansible.html`

- ansible-config: `https://docs.ansible.com/ansible/latest/cli/ansible-config.html`

- ansible-console: `https://docs.ansible.com/ansible/latest/cli/ansible-console.html`

- ansible-doc: `https://docs.ansible.com/ansible/latest/cli/ansible-doc.html`

- ansible-inventory: `https://docs.ansible.com/ansible/latest/cli/ansible-inventory.html`

- ansible-playbook: `https://docs.ansible.com/ansible/latest/cli/ansible-playbook.html`

- ansible-pull: `https://docs.ansible.com/ansible/latest/cli/ansible-pull.html`

- ansible-vault: `https://docs.ansible.com/ansible/latest/cli/ansible-vault.html`

- ansible-inventory-grapher: `https://github.com/willthames/ansible-inventory-grapher`

# Part 2: Deploying Applications

Now that you understand Ansible's basics, it's time to put that knowledge into practice. In this part, we will focus on deploying applications using Ansible playbooks. From setting up a LAMP stack to deploying WordPress and targeting multiple distributions, you will gain hands-on experience automating application deployments. We will also explore how Ansible can manage Windows-based servers, expanding your automation capabilities.

This part has the following chapters:

- *Chapter 4, Deploying a LAMP Stack*
- *Chapter 5, Deploying WordPress*
- *Chapter 6, Targeting Multiple Distributions*
- *Chapter 7, Ansible Windows Modules*

# 4
# Deploying a LAMP Stack

This chapter will look at deploying a complete LAMP stack using the various core modules that ship with Ansible. We will target the local Multipass virtual machine we first used in *Chapter 1, Installing and Running Ansible*.

We will discuss the following:

- The playbook layout – how our playbook is going to be structured
- Linux – preparing the Linux server
- Apache – installing and configuring Apache
- MariaDB – installing and configuring MariaDB
- PHP – installing and configuring PHP

This chapter covers the following topics:

- The playbook structure
- The LAMP stack
- The LAMP playbook

Before we start writing the playbook, we will discuss the structure we will use after we briefly discuss what we need for the chapter.

## Technical requirements

We will again use the local Multipass virtual machine we launched in the previous chapters. As we will be installing all the elements of a LAMP stack on the virtual machine, your Multipass virtual machine will need to be able to download packages from the internet; in all, there is around 500 MB of packages and configuration to download.

You can find a complete copy of the playbook in the repository accompanying this book at `https://github.com/PacktPublishing/Learn-Ansible-Second-Edition/tree/main/Chapter04`.

## The playbook structure

In *Chapter 1*, *Installing and Running Ansible*, the playbooks we ran were as basic as possible. They have been in a single file, accompanied by a host inventory file, and, if required, a template file. Then, in *Chapter 2*, *Exploring Ansible Galaxy*, we extended our playbook files to include roles rather than putting all our tasks, handlers, and variables into one file.

As you can see from the following layout, there are several folders and files:

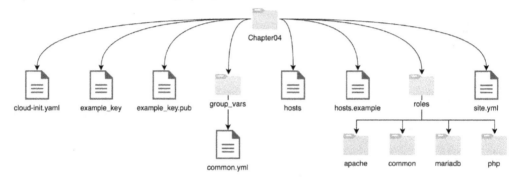

Figure 4.1 – The folder structure we will use for our playbook

While there is a copy of the structure in the repository, let's work on creating the structure and discuss each item as we create it. The first folder we need to create is our top-level folder. This is the folder that will contain our playbook folders and files:

```
$ mkdir Chapter04
$ cd Chapter04
```

The next folder we are going to create is one called `group_vars`. This will contain the variable files used in our playbook. For now, we are going to create a single variable file called `common.yml`:

```
$ mkdir group_vars
$ touch group_vars/common.yml
```

Next, we are going to create two files – our host inventory file, which we will name `hosts`, and our master playbook, which is typically called `site.yml`:

```
$ touch production
$ touch site.yml
```

The final folder we are going to create manually is called `roles`. Here, we are going to use the `ansible-galaxy` command, which we learned about in *Chapter 2, Exploring Ansible Galaxy*, to create a role called `common`. To do this, we use the following commands:

```
$ mkdir roles
$ ansible-galaxy role init roles/common
```

This should create all the files needed to start writing the `common` role.

The `cloud-init.yaml`, `example_key`, `example_key.pub`, and `hosts.example` files are all lifted straight from *Chapter 1, Installing and Running Ansible*, and *Chapter 2, Exploring Ansible Galaxy*, so we will not cover them again in this chapter.

> **Note**
>
> While we will work through each of the files individually in this and the following sections, a complete copy of the playbook is available in the accompanying GitHub repository.

Let's look at each of the four roles in our playbook and install and configure our LAMP stack.

## The LAMP stack

The LAMP stack is the term used to describe an all-in-one web and database server. Typically, the components are as follows:

- **Linux** is the underlying operating system; in our case, we will use Ubuntu 22.04
- **Apache** is the web server element of the stack
- **MariaDB** is what we will use as the database component of the stack; typically, it is based on **MySQL**, which could also be used
- **PHP** is the dynamic language used by the web server to generate content

A common variation of the **LAMP** stack is called **LEMP**; this replaces **Apache** with **NGINX**, which is pronounced *engine-x*, hence the *E* rather than the *N*.

We are going to look at creating roles to deal with these components; these are as follows:

- `common`: This role will prepare our Ubuntu server, installing any supporting packages and services we need
- `apache`: This role will install the Apache web server and configure a default virtual host
- `mariadb`: This role will not only install MariaDB but also secure the installation and create a default database and user, as well as optionally download and import a database to use

- php: This role will install PHP and configure a set of common PHP modules, and if we set the option to a database admin tool written in PHP, we can interact with our test database via the browser

Let us begin by looking at the common role.

## The common role

In the previous section of this chapter, we used the `ansible-galaxy role init` command to create the common role. This creates several folders and files; as discussed in *Chapter 2, Exploring Ansible Galaxy*, we will not go into any detail here but instead dive straight into the role itself.

Let's make a start by adding some tasks.

### Updating installed packages

First of all, let's update our server by adding the following to the beginning of the `roles/common/tasks/main.yml` file:

```
- name: "Update apt cache and upgrade packages"
  ansible.builtin.apt:
    name: "*"
    state: "latest"
    update_cache: true
```

You will notice a difference from when we last used the `ansible.builtin.apt module` to update all the installed packages.

We now start the task using the `name` key; this will print out the content of the value we assigned to the name key when the playbook runs, which will give us a better idea of what is going on during the playbook run, rather than just printing the name of the module that is executed.

### Installing common packages

Now that we have updated the installed packages, let's install the packages we want to install on all the Linux servers we will target with the Playbook:

```
- name: "Install common packages"
  ansible.builtin.apt:
    state: "present"
    pkg: "{{ common_packages }}"
```

As you can see, we again use the `ansible.builtin.apt` module, and we have added a descriptive name for the task. Rather than providing a list of packages in the task, we use a variable called common_packages, which is defined in the `roles/common/defaults/main.yml` file as follows:

```
common_packages:
    - "ntp"
    - "sntp"
    - "ntp-doc"
    - "vim"
    - "git"
    - "unzip"
```

As you can see, we install `ntp`, `sntp`, and `ntp-doc`; we will configure `ntp` shortly. Next, we install `vim`, `git`, and `unzip`, as they are always helpful to have installed on a server.

Another thing that you may have noticed is that we pass a list of packages using {{ common_packages }} to the `pkg` key in the `ansible.builtin.apt` module, resulting in the module looping through the list of packages we pass in and installing them all in one go, rather than having to call the module to install each package individually.

### Configuring Network Time Protocol (NTP)

Next, we copy the `ntp.conf` file from the `templates` folder, adding the list of NTP servers as we have done in the previous chapters, and then informing Ansible to restart NTP whenever the configuration file changes.

### Creating a key, group, and user

In the `roles/common/defaults/main.yml` file, the following variable is defined:

```
users:
    - {
        name: "lamp",
        group: "lamp",
        state: "present",
        key: "/tmp/id_ssh_lamp_rsa",
    }
```

This is slightly different from the variables we have used so far, as it is a single variable called `users`, which is made up of a single item, and that item contains the `name`, `group`, `state`, and `key` key-value pairs.

Because we are using items, we need to change our approach to how we use the variables within the task, the first of which in `roles/common/tasks/main.yml` creates an OpenSSH key pair; if one doesn't already exist, we need to save it at the path that is defined in the key key-value pair:

```
- name: "Generate a ssh keypair"
  community.crypto.openssh_keypair:
    path: "{{ item.key }}"
  with_items: "{{ users }}"
  delegate_to: "localhost"
  become: false
```

Working through the task, you can see that we use the `community.crypto.openssh_keypair` module, in which we pass just one value, which is the path to the file where we would like our OpenSSH key stored.

As you can see, we use the `{{ item.key }}` variable to enter the path, but we do not define that the variable is called `users` here; instead, we use the `with_items` option and pass in the `{{ users }}` variable here.

While we only pass one item in this example, you could take this approach to execute a single task multiple times – for example, if our variable looked like this:

```
users:
  - {
      name: "lamp",
      group: "lamp",
      state: "present",
      key: "/tmp/id_ssh_lamp_rsa",
    }
  - {
      name: "user2",
      group: "lamp",
      state: "present",
      key: "/tmp/id_ssh_user2_rsa",
    }
```

Then, when the task is executed, it would create two OpenSSH keys, and the subsequent tasks, which we will get into in a moment, would create a single group called `lamp` and then two users, `lamp` and `user2`.

Back to the task at hand – you will notice that we have defined two other options, `delegate_to` and `become`.

If we were to run the community.crypto.openssh_keypair module without defining delegate_to, then the module will be executed on the remote host, which is not what we want to happen in this case, as we want a copy of the private and public portions of the OpenSSH key on our local machine. Therefore, by using localhost as the value in the delegate_to option, we tell Ansible to run this task locally.

The next option, become, tells Ansible not to become an escalated user using the sudo command, which is the default action for all the hosts we have defined at the top of our main site.yml playbook file – this is because we want the community.crypto.openssh_keypair module to run as the user you are logged in as, rather than your local machine's root user.

The logic for this task, minus the delegate_to and become options, as we want the remainder of the tasks to be executed against the target machine, is followed through to the remaining tasks in the role, starting with creating the group by executing the ansible. builtin.group module:

```
- name: "Add group for our users"
  ansible.builtin.group:
    name: "{{ item.group }}"
    state: "{{ item.state }}"
  with_items: "{{ users }}"
```

Once the group has been created, we can then add the user using ansible.builtin.user, or users if we have defined more than one item in the users variable:

```
- name: "Add users to our group"
  ansible.builtin.user:
    name: "{{ item.name }}"
    group: "{{ item.group }}"
    comment: "{{ item.name }}"
    state: "{{ item.state }}"
  with_items: "{{ users }}"
```

The final task in the role takes the public portion of the OpenSSH key, which we generated earlier, and adds the contents to the user(s) created during the previous task, using the ansible.builtin. authorized_key module:

```
- name: "Add keys to our users"
  ansible.posix.authorized_key:
    user: "{{ item.name }}"
    key: "{{ lookup('file', item.key + '.pub') }}"
  with_items: "{{ users }}"
```

You may have noticed that the value we pass for the key option is new to us; this uses the lookup plugin to read the file's contents at the item.key path with .pub appended to the end, meaning, in our case, it reads the contents of the file at /tmp/id_ssh_lamp_rsa.pub. This file is the public portion of the OpenSSH key-pair, which was created when we executed the "generate a ssh keypair" task earlier in the role.

The lookup plugin is designed to be executed locally, so in this case, we do not need to use the delegate_to and become options, as we want the task to be executed on the target host because that is where our user has been created, but we want to populate the /home/lamp/.ssh/authorized_key file on the remote host with the contents of the /tmp/id_ssh_lamp_rsa.pub file that we have on our local host.

That concludes the tasks in the common role; before we move on to the next role, which will install and configure apache, you should know one more thing.

The "generate a ssh keypair" task will not overwrite any existing key-pairs when executed, meaning the first time you run the role and no files exist at /tmp/id_ssh_lamp_rsa and /tmp/id_ssh_lamp_rsa.pub, the key-pair will be created, and on subsequent Playbook runs, as the files now exist, the task will return an **OK**, as there is nothing for it to do, as a valid key-pair is already at the location, and we ask the community.crypto.openssh_keypair module to create the key-pair.

## The Apache role

Once the common role has finished running our remote host, we will be ready to install and configure the Apache web server.

### Installing the Apache packages

The first task in roles/apache/tasks/main.yml installs the packages we need to run the Apache web server; it uses the ansible.builtin.apt module and looks like the following:

```
- name: "Install apache packages"
  ansible.builtin.apt:
    state: "present"
    pkg: "{{ apache_packages }}"
```

As you can see, it calls a variable called {{ apache_packages }}, which is defined in roles/apache/defaults/main.yml as follows:

```
apache_packages:
  - "apache2"
  - "apache2-ssl-dev"
  - "ca-certificates"
  - "openssl"
```

As we learned when we walked through the common role, this will install the four packages defined in the variable.

Once Apache has been installed, which is a single task, we can now progress to configuring our Apache installation.

## Configuring Apache

The first task when configuring Apache is to take the user that was created when the `common` role was run and add them to the Apache group; to do this, we run the following task:

```
- name: "Add user to apache group"
  ansible.builtin.user:
    name: "{{ item.name }}"
    groups: "{{ apache_group }}"
    append: true
  with_items: "{{ users }}"
```

This takes the `{{ users }}` variable from the previous role and loops over the items defined in the variable, adding the user to the group defined under the `{{ apache_group }}` variable in the `roles/apache/defaults/main.yml` file. A full list of the variables defined to configure Apache, which we will use throughout the next few tasks, is as follows:

```
apache_group: "www-data"
web_root: "web"
document_root: "/home/{{ users.0.name }}/{{ web_root }}"
index_file: index.html
vhost_path: "/etc/apache2/sites-enabled/"
vhost_default_file: "000-default.conf"
vhost_our_file: "vhost.conf"
```

You may have noticed that the value of the `document_root` variable is a little different from the ones we have used so far; there'll be more on that in a moment.

The next task creates a folder within the users directly, which we will use to store the files served via Apache:

```
- name: "Create the document root for our website"
  ansible.builtin.file:
    dest: "{{ document_root }}"
    state: directory
    mode: "0755"
    owner: "{{ users.0.name }}"
    group: "{{ apache_group }}"
```

As you can see, we use `{{ users.0.name }}` as we did for the `document_root` variable value; why is this?

As we know, the common role only creates a single user; we can't simply use {{ users.name }}, as the name key exists within an item within the variable, so using {{ users.name }} would result in an error, stating that the variable can't be found.

Because of this, we can reference the first item in the list of items by using its position within the list, which, because Ansible counts from zero, will be 0 rather than 1.

Using the values that we have defined in the defaults for the common and apache roles, this task will create a folder at /home/lamp/web/; the lamp user would own the folder and would be assigned to the www-data group, which is the group the Apache process will run as.

The next task will ensure the correct read, write, and execute permissions are set on the /home/ lamp/ folder:

```
- name: "Set the permissions on the user folder"
  ansible.builtin.file:
    dest: /home/{{ users.0.name }}/
    state: directory
    mode: "0755"
    owner: "{{ users.0.name }}"
```

That task concludes configuring the folder structure needed to serve our web pages; now, it is time to configure Apache itself.

The first thing we need to do is remove the default virtual host configuration file; to do this, we will execute the following task:

```
- name: "Remove the apache default vhost config"
  ansible.builtin.file:
    path: "{{ vhost_path }}{{ vhost_default_file }}"
    state: absent
  notify: "Restart apache2"
```

This uses the ansible.builtin.file module to set the state of the file defined by {{ vhost_default_file }} in the {{ vhost_path }} folder to absent, which means, if the file exists, remove it.

It also uses notify to call the "Restart apache2" handler, which is defined as the following task in the roles/apache/handlers/main.yml file:

```
- name: "Restart apache2"
  ansible.builtin.service:
    name: "apache2"
    state: "restarted"
    enabled: true
```

Once the default file has been removed, we can add our virtual host configuration file.

The template for this virtual host configuration file can be found at `roles/apache/templates/vhost.conf.j2`, and it contains the following:

```
# {{ ansible_managed }}
<VirtualHost *:80>
  ServerName {{ ansible_hostname }}
  DocumentRoot {{ document_root }}
  DirectoryIndex {{ index_file }}
  <Directory {{ document_root }}>
    AllowOverride All
    Require all granted
  </Directory>
</VirtualHost>
```

When loaded, this configuration file serves the contents of the `{{ document_root }}` folder when someone visits the site's URL in their browser.

The task to deploy this template file to the remote host looks like the following:

```
- name: "Copy the our vhost.conf to the sites-enabled folder"
  ansible.builtin.template:
    src: vhost.conf.j2
    dest: "{{ vhost_path }}{{ vhost_our_file }}"
    mode: "0644"
  notify: "Restart apache2"
```

As you can see, this also calls the `"Restart apache2"` handler if there are any changes to the file.

With Apache now configured, there is one final task.

### Optionally copying an index.html file

The final task in this role uses the following `variables` block:

```
html_deploy: true
html_heading: "Success !!!"
html_body: |
  This HTML page has been deployed using Ansible to <b>{{ ansible_host
  }}</b>.<br>
  The user is <b>{{ users.0.name }}</b> who is in the <b>{{ apache_
  group }}</b> group.<br>
  The weboot is <b>{{ document_root }}</b>, the default index file is
  <b>{{ index_file }}</b>.<br>
```

As you can see, it contains a heading and some HTML code for the body; these variables are used by the following task:

```
- name: "Copy the test HTML page to the document root"
  ansible.builtin.template:
    src: index.html.j2
    dest: "{{ document_root }}/index.html"
    mode: "0644"
    owner: "{{ users.0.name }}"
    group: "{{ apache_group }}"
  when: html_deploy
```

This uses a template that can be found at `roles/apache/templates/index.html.j2` and looks like the following:

```
<!--{{ ansible_managed }}-->
<!doctype html>
<title>{{ html_heading }}</title>
<style>
  body { text-align: center; padding: 150px; }
  h1 { font-size: 50px; }
  body { font: 20px Helvetica, sans-serif; color: #333; }
  article { display: block; text-align: left; width: 650px; margin: 0
auto; }
</style>
<article>
    <h1>{{ html_heading }}</h1>
    <div>
        <p>{{ html_body }}</p>
    </div>
</article>
```

However, the task is only called if the `html_deploy` variable is set to `true`; this is managed by the following statement at the end of the task:

```
when: html_deploy
```

So, if, for any reason, the `html_deploy` variable is not equal to `true`, then the task will be skipped when the playbook is executed.

That's all we need to do to install and configure Apache; let us now look at installing the M in LAMP and review the role to install and configure MariaDB.

# The MariaDB role

Of the four roles we cover in this chapter, this, the MariaDB one, is the most complicated, as it installs MariaDB, configures it, and optionally downloads and imports a sample database.

Let's start by covering the installation.

## Installing MariaDB

You may have started to spot a trend in the roles; the tasks always start with installing a few packages, and MariaDB is no different.

The task from `roles/mariadb/tasks/main.yml` is as follows:

```
- name: "Install mariadb packages"
  ansible.builtin.apt:
    state: "present"
    pkg: "{{ mariadb_packages }}"
```

The `mariadb_packages` variable in `roles/mariadb/defaults/main.yml` looks like the following:

```
mariadb_packages:
  - "mariadb-server"
  - "mariadb-client"
  - "python3-pymysql"
```

As you can see, we installed the MariaDB client and server. Also, we installed the `python3-pymysql` package; this is required for the tasks that need to interact with MariaDB once it is installed to function. Without it, Ansible cannot establish a connection to and interact with our MariaDB server.

Once the packages have been installed, we need to start the MariaDB server by using the following task:

```
- name: "Start mariadb"
  ansible.builtin.service:
    name: mariadb
    state: started
    enabled: true
```

You might be thinking, why aren't we using a handler as we have done for previous tasks? Well, handlers are only called once the playbook execution has been completed and Ansible knows all the services that need to be restarted.

However, in this case, we need to interact with the MariaDB service to be able to configure it as part of the playbook run, so rather than using a handler, we just start the service as a task using the same block we would use as the handler.

Now that MariaDB is installed and started, we can start the configuration.

### Configuring MariaDB

Before we dive into the tasks, quickly look at the variables in `roles/mariadb/defaults/main.yml`, which will be used to configure our MariaDB server:

```
mariadb_root_username: "root"
mariadb_root_password: "Pa55W0rd123"
mariadb_hosts:
  - "127.0.0.1"
  - "::1"
  - "{{ ansible_nodename }}"
  - "%"
  - "localhost"
```

Now that we know what variables we will use, it's time to work through the configuration, which is a little complex due to the default way that MariaDB is configured when it starts immediately after installation.

By default, MariaDB starts with no password in place, meaning that anyone can connect to the database as the root user, which is not ideal, so the first thing we need to do is to secure our installation by setting the root password.

That sounds easy enough, you might be thinking to yourself.

Technically, it is; however, if the playbook were to be run a second time, meaning that there is now a password set, then the task we are about to define, which sets the initial password, will error, as we need to configure the task not to use a password. Once the password has been set, the server will only accept a connection using the already set password.

We also need to consider that once a password has been configured, we need to use that password each time we need to connect to the MariaDB server – so we need an easy way to ensure we can connect smoothly once the password has been set.

Luckily, there is a function built into MariaDB and MySQL that allows you to put your credentials into a file on the server; the file should be placed in the home directory of the user you are logged in as. Once in place, each time you attempt to connect to the database server using that user, the database client will read the file and connect you, without you having to type the credentials – this file should be called `~/.my.cnf` (the `~/` part is a shortcut for the user's home folder).

For our scenario, this works because we can check for the presence of the `~/.my.cnf` file, and if it is not there, then it will be safe to assume that the password has not been configured yet.

The task that checks for the presence of the file is as follows:

```
- name: "Check to see if the ~/.my.cnf file exists"
  ansible.builtin.stat:
    path: ~/.my.cnf
  register: mycnf
```

This uses the `ansible.builtin.stat` module to check for the file and then uses the `register` option to register a runtime variable, called `mycnf`.

Now that we have a dynamically registered variable that contains details on whether the `~/.my.cnf` file exists on the remote host's filesystem or not, we can now proceed with changing the password or skip the task if the `~/.my.cnf` file is present.

Ansible has several built-in modules to interact with MySQL and MariaDB; the one we will use here is `ansible.builtin.mysql_user`:

```
- name: "Change mysql root password if we need to"
  community.mysql.mysql_user:
    name: "{{ mariadb_root_username }}"
    host: "{{ item }}"
    password: "{{ mariadb_root_password }}"
    check_implicit_admin: "yes"
    priv: "*.*:ALL,GRANT"
    login_user: "{{ mariadb_root_username }}"
    login_unix_socket: /var/run/mysqld/mysqld.sock
  with_items: "{{ mariadb_hosts }}"
  when: not mycnf.stat.exists
```

In the task, we instruct Ansible to set the password for the user defined in the `{{ mariadb_root_username }}` variable to the password stored in the `{{ mariadb_root_password }}` variable, giving the user full admin access to all the databases across all possible host combinations, which are defined in the `{{ mariadb_hosts }}`, which we loop over using the `with_items` function.

When logging in to do this, Ansible should use the `{{ mariadb_root_username }}` username and connect over a Unix socket, which can be found at `/var/run/mysqld/mysqld.sock`; this means we don't have to establish a network connection to interact with the database because, if we did, Ansible wouldn't be able to connect, as it can't send a blank password.

Finally, only run this task when the `mycnf.stat.exists` variable is equal to `false`.

Now that we have set the actual password and secured the MariaDB installation, we need to create the `~/.my.cnf` file to carry on with the configuration.

To do this, we will again use a template, which can be found at `roles/mariadb/templates/my.cnf.j2`. This template looks like the following:

```
# {{ ansible_managed }}
[client]
user='{{ mariadb_root_username }}'
password='{{ mariadb_root_password }}'
```

As you can see, it contains the username and password needed to connect to the database server.

Because the file contains credentials, when the task creates the file on the server, we need to ensure that the file can only be read and written to by the root user, by setting the read, write, and execute permissions of the file as it is created:

```
- name: "Set up ~/.my.cnf file"
  ansible.builtin.template:
    src: "my.cnf.j2"
    dest: "~/.my.cnf"
    mode: "0600"
```

Now that we have the `~/.my.cnf` file on the remote host, we can progress with securing our MariaDB installation; the subsequent task removes the anonymous user, again looping through the hosts that user could be associated with:

```
- name: "Delete anonymous MySQL user"
  community.mysql.mysql_user:
    user: ""
    host: "{{ item }}"
    state: absent
  with_items: "{{ mariadb_hosts }}"
```

The final task that deals with securing our MariaDB installation removes the default `test` database:

```
- name: "Remove the MySQL test database"
  community.mysql.mysql_db:
    db: "test"
    state: "absent"
```

The remainder of the tasks in the role, such as copying the `index.html` file in the `apache` role, are optional, so let's review those tasks now.

### Downloading and importing the example database

There is one more block of variables in `roles/mariadb/defaults/main.yml`; these deal with downloading and importing an example database. There are a lot of keys in the `mariadb_sample_database` variable, starting with the flag to enable the option, the URL of the file to download, and the path to save it to:

```
mariadb_sample_database:
  create_database: true
  source_url: "https://github.com/russmckendrick/test_db/archive/
master.zip"
  path: "/tmp/test_db-master"
```

Next, we have the name of the example database being created as well as the username and password to use for the new database:

```
db_name: "employees"
db_user: "employees"
db_password: "employees"
```

Finally, there is a list of the files that need to be imported. The first two files contain the schema:

```
dump_files:
  - "employees.sql"
  - "show_elapsed.sql"
```

The remaining files contain the actual data to be loaded:

```
  - "load_departments.dump"
  - "load_employees.dump"
  - "load_dept_emp.dump"
  - "load_dept_manager.dump"
  - "load_titles.dump"
  - "load_salaries1.dump"
  - "load_salaries2.dump"
  - "load_salaries3.dump"
```

Now that we know what variables are defined, we can work through the remaining tasks, the first of which downloads and unarchives the ZIP file that contains the example database files:

```
- name: "Download and unarchive the sample database data"
  ansible.builtin.unarchive:
    src: "{{ mariadb_sample_database.source_url }}"
    dest: /tmp
    remote_src: "yes"
  when: mariadb_sample_database.create_database
```

As you can see, the `ansible.builtin.unarchive` module allows you to download and unarchive the file, meaning we can do everything we need in a single task. Also, we only run the when task when the `mariadb_sample_database.create_database` equals `true`. We will do this for the remainder of the tasks and even expand upon the when statement toward the end of the role.

The next task creates the example database:

```
- name: "Create the sample database"
  community.mysql.mysql_db:
    db: "{{ mariadb_sample_database.db_name }}"
    state: present
  when: mariadb_sample_database.create_database
```

Once the database has been created, we can run a task that creates the user and assigns permissions to the newly created user to access the database we just added:

```
- name: "Create the user for the sample database"
  community.mysql.mysql_user:
    name: "{{ mariadb_sample_database.db_user }}"
    password: "{{ mariadb_sample_database.db_password }}"
    priv: "{{ mariadb_sample_database.db_name }}.*:ALL"
    state: present
  with_items: "{{ mariadb_hosts }}"
  when: mariadb_sample_database.create_database
```

We are now down to the final two tasks, and here is where we need to add a little more logic to our playbook to ensure that we only import the example data once; if we don't have the logic in place, we can run into all sorts of problems if the playbook is rerun and could risk data being overwritten or duplicate data being inserted if the import task is allowed to run again.

As the databases are stored on the host's filesystem, we can use the same logic that we used to check for the presence of the `~/.my.cnf` file, but this time, we check for a database file:

```
- name: "Check to see if we need to import the sample database dumps"
  ansible.builtin.stat:
    path: /var/lib/mysql/{{ mariadb_sample_database.db_name }}/{{
mariadb_sample_database.db_name }}.frm
  register: db_imported
  when: mariadb_sample_database.create_database
```

We register a variable called `db_imported`, which we will use with the when condition of the next and final task; this is the one that loops through `mariadb_sample_database.dump_files` and imports the databases:

```
- name: "Import the sample database"
  community.mysql.mysql_db:
```

```
    name: "{{ mariadb_sample_database.db_name }}"
    state: import
    target: "{{ mariadb_sample_database.path }}/{{ item }}"
  with_items: "{{ mariadb_sample_database.dump_files }}"
  when: db_imported is defined and not db_imported.stat.exists
```

We have changed the when condition slightly here; rather than referencing `mariadb_sample_database.create_database`, we only use `db_imported`.

The first part ensures that the playbook doesn't error if we decide not to import the database by setting `mariadb_sample_database.create_database` to `false`, as `db_imported` can only be defined if `mariadb_sample_database.create_database` is set to `true`, as the task that sets the `db_imported` variable is only ever executed when that condition is met.

As you can also see, we use `and`, thus adding a second condition to the when statement; this means that the task will only be executed if `db_imported is defined` and `not db_imported.stat.exists` are both met.

That final task brings us to the end of the MariaDB role and leaves us with one role to work through – the PHP role.

## The PHP role

This, our final role, installs PHP, optionally copies a PHP Info file along with it, and installs a database management interface written in PHP, called Adminer, so that we can access the database server we used in the previous role.

### Installing the PHP packages

It should come as no surprise to you that the first task executed in the PHP role installs the packages needed for us to run PHP.

The full list of packages is defined in the `roles/php/default/main.yml` file, as follows:

```
php_packages:
  - "php"
  - "php-cli"
  - "php-curl"
  - "php-gd"
  - "php-intl"
  - "php-mbstring"
  - "php-mysql"
  - "php-soap"
  - "php-xml"
  - "php-xmlrpc"
```

```
  - "php-zip"
  - "libapache2-mod-php"
```

The task itself looks familiar:

```
- name: "Install php packages"
  ansible.builtin.apt:
    state: "present"
    pkg: "{{ php_packages }}"
  notify: "Restart apache2"
```

The thing to note is that we restart Apache once PHP is installed because we run PHP as an Apache module. So, once installed, Apache needs to be restarted to load in the module and enable PHP on our Apache web server.

That's it. PHP is installed, and Apache asks to be restarted; everything from here is optional.

### Copying the PHP Info file

The next task is a simple one that copies `roles/php/files/info.php` to the web root of the server if the `php_info` variable is set to `true`:

```
- name: "Copy the PHP info to the document root"
  ansible.builtin.copy:
    src: info.php
    dest: "{{ document_root }}/info.php"
    mode: "0755"
    owner: "{{ users.0.name }}"
    group: "{{ apache_group }}"
  when: php_info
```

The only difference is that we copy the file from our local host to the remote one with this task – we do not use the `ansible.builtin.template` module this time but instead, the `ansible.builtin.copy` one. This is because `info.php` is made up of three lines of code, none of which we need to update based on the environment or any variables we set.

### Installing and configuring Adminer

The variables for the remaining tasks in the `roles/php/default/main.yml` file look like the following:

```
adminer:
  install: true
  path: "/usr/share/adminer"
  download: "https://github.com/vrana/adminer/releases/download/
v4.8.1/adminer-4.8.1-mysql.php"
```

They define where to download the file from and where on the remote to download it to, which is where the first of the three tasks comes in, as this creates the folder on the remote virtual machines filesystem for us to download Adminer to:

```
- name: "Create the document root for adminer"
  ansible.builtin.file:
    dest: "{{ adminer.path }}"
    state: directory
    mode: "0755"
  when: adminer.install
```

Once we have the download target folder created, we can download Adminer itself:

```
- name: "Download adminer"
  ansible.builtin.get_url:
    url: "{{ adminer.download }}"
    dest: "{{ adminer.path }}/index.php"
    mode: "0755"
  when: adminer.install
```

As you may have spotted from the download URL and destination, Adminer is a single PHP file that we save as index.php. So, how will we access Adminer via our Apache web server?

Well, to do that, we need to copy across another virtual host configuration file:

```
- name: "Copy the adminer.conf to sites-enabled folder"
  ansible.builtin.template:
    src: adminer.conf.j2
    dest: "{{ vhost_path }}adminer.conf"
    mode: "0755"
  when: adminer.install
  notify: "Restart apache2"
```

As you can see, this renders and copies across roles/php/templates/adminer.conf.j2 to adminer.conf, the site-enabled folder on our remote host, and instructs the Apache service to restart to load the newly added configuration.

The adminer.conf.j2 file contains the following:

```
# {{ ansible_managed }}
Alias /adminer "{{ adminer.path }}"
  <Directory "{{ adminer.path }}">
    DirectoryIndex index.php
    AllowOverride All
    Require all granted
  </Directory>
```

This tells Apache that whenever someone visits `http://someurl/adminer/`, the Adminer `index.php` file should be served.

With that task covered, we have completed the walk-through of the four roles that go into installing and configuring our LAMP stack, and now it is time to review and execute the playbook itself.

## The LAMP playbook

As mentioned at the start of this chapter when we discussed the playbook structure, the main playbook file is called `site.yml`, which contains the following:

```
---


- name: "Install LAMP stack"
  hosts: ansible_hosts
  gather_facts: true
  become: true
  become_method: ansible.builtin.sudo

  vars_files:
    - group_vars/common.yml

  roles:
    - common
    - apache
    - mariadb
    - php
```

As you can see, it calls the four roles we have already walked through and also loads a `variables` file from `group_vars/common.yml`; this file contains an override for `html_body`, which is configured in `roles/apache/defaults/main.yml` and looks like the following:

```
html_body: |
  This HTML page has been deployed using Ansible to <b>{{ ansible_
nodename }}</b>.<br>
  The user is <b>{{ users.0.name }}</b> who is in the <b>{{ apache_
group }}</b> group.<br>
  The weboot is <b>{{ document_root }}</b>, the default index file is
<b>{{ index_file }}</b>.<br><br>
  You can access a <a href="/info.php">PHP Info file</a> or <a href="/
adminer/">Adminer</a>.
```

This means that when we run the playbook, the `index.hml` page will have links to `info.php` and the `/adminer` URL to access the additional content easily.

> **Note**
>
> The Chapter04 folder in the GitHub repo that accompanies this title contains the example hosts file and keys to launch a local virtual machine using Multipass. If you are following along, refer to the instructions in *Chapter 1, Installing and Running Ansible,* for how to launch the virtual machine and prepare your own hosts file.

So, without further ado, let's run the playbook:

```
$ ansible-playbook -i hosts site.yml
```

On the first run, this should give us some output that looks like the following:

```
PLAY [ansible_hosts]
TASK [Gathering Facts]
ok: [ansiblevm]
TASK [roles/common : update apt cache and upgrade packages]
ok: [ansiblevm]
…. lots of other output here ….
RUNNING HANDLER [roles/apache : restart apache2]
changed: [ansiblevm]
PLAY RECAP
ansiblevm : ok=34     changed=26
unreachable=0     failed=0     skipped=0     rescued=0     ignored=0
```

As you can see, the Playbook has made 26 changes to the target virtual machine.

Let's run the playbook a second time:

```
$ ansible-playbook -i hosts site.yml
```

Then, in the play recap, you should see that some tasks were skipped:

```
PLAY RECAP
ansiblevm       : ok=30     changed=0
unreachable=0     failed=0     skipped=2     rescued=0     ignored=0
```

As expected, one of those tasks was updating the root password for the database user:

```
TASK [roles/mariadb : change mysql root password if we need to]
skipping: [ansiblevm] => (item=127.0.0.1)
skipping: [ansiblevm] => (item=::1)
skipping: [ansiblevm] => (item=ansiblevm)
skipping: [ansiblevm] => (item=%)
skipping: [ansiblevm] => (item=localhost)
skipping: [ansiblevm]
```

The second task that is skipped is importing the database files:

```
TASK [roles/mariadb : import the sample database]
skipping: [ansiblevm] => (item=employees.sql)
skipping: [ansiblevm] => (item=show_elapsed.sql)
skipping: [ansiblevm] => (item=load_departments.dump)
skipping: [ansiblevm] => (item=load_employees.dump)
skipping: [ansiblevm] => (item=load_dept_emp.dump)
skipping: [ansiblevm] => (item=load_dept_manager.dump)
skipping: [ansiblevm] => (item=load_titles.dump)
skipping: [ansiblevm] => (item=load_salaries1.dump)
skipping: [ansiblevm] => (item=load_salaries2.dump)
skipping: [ansiblevm] => (item=load_salaries3.dump)
skipping: [ansiblevm]
```

Both are to be expected, as that is how we configured the tasks to respond on subsequent Playbook runs.

Now, if you open your browser and enter http:// and then the name of your Ansible host (for me, this was http://192.168.64.20.nip.io; I suspect yours will be different, so the link will likely not work), then you should be greeted by the index.html page that Ansible generated:

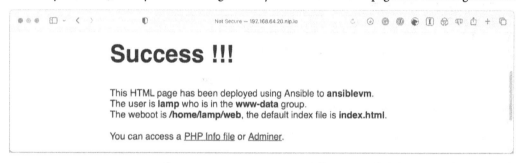

Figure 4.2 – Success !!! – viewing the index.html page

Clicking on the link for the PHP Info file should take you to something like `http://192.168.64.20.nip.io/info.php`, which will display information on your PHP installation:

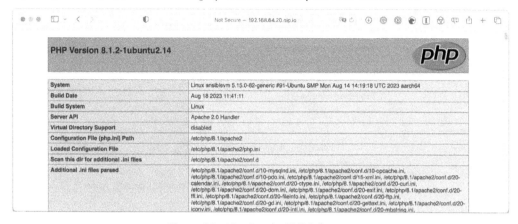

Figure 4.3 – Viewing the PHP Info page

The final link to click is the one for Adminer; clicking it will take you to `http://192.168.64.20.nip.io/adminer/`, which will prompt you to log in:

Figure 4.4 – The Adminer login page

To log in, use the following credentials:

- **Username**: `root`
- **Password**: `Pa55W0rd123`
- **Database**: `employees`

Once logged in, you will be taken straight to an overview of the **employees** database:

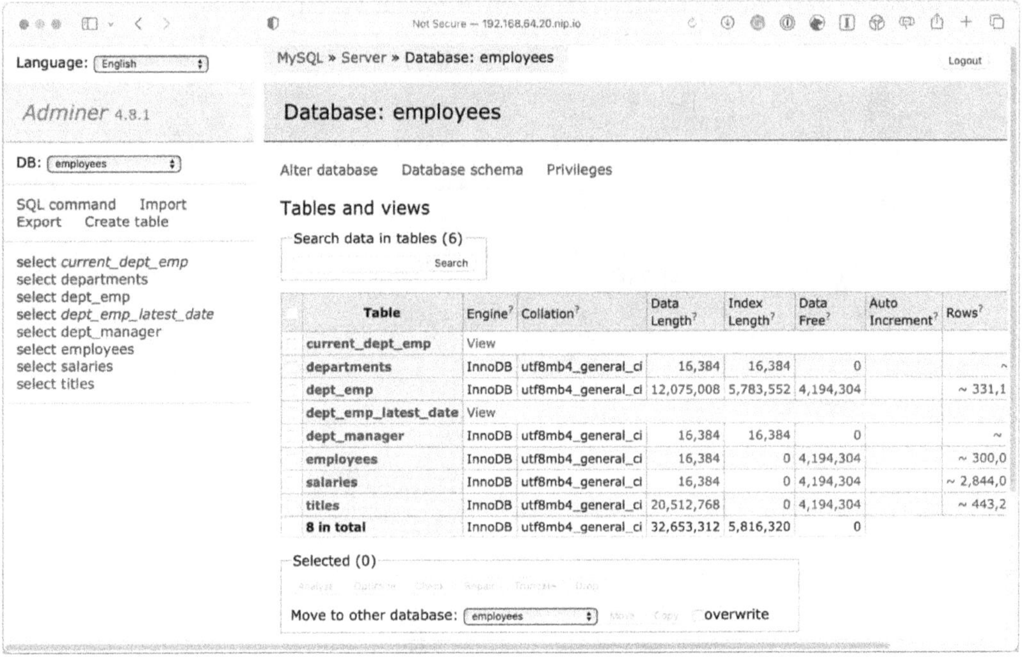

Figure 4.5 – The employees database overview

Feel free to click around, and once you have finished, ensure that you terminate the Multipass virtual machine; instructions on how to do this can be found at the end of *Chapter 1*, *Installing and Running Ansible*.

## Summary

In this chapter, we worked through writing a playbook that installs a LAMP stack on our Multipass virtual machine. We created four roles, one for each element of the stack, and within each of the roles, we built in a bit of logic that can be overridden to deploy additional elements, such as test HTML and PHP pages, and we also built in the option to create a test database that contains over 40,000 records.

So far, we installed some basic packages. In the next chapter, we will write a playbook that installs, configures, and maintains a WordPress installation.

This updated playbook will reuse some of the elements from the roles we covered in this chapter and make some improvements, as some of the elements we covered in this chapter were a little too simplistic. The biggest change is that we will not use a hardcoded password for the database instance moving forward.

# Further reading

You can find the project pages for the third-party tools covered throughout the chapter at the following URLs:

- **Apache**: `https://httpd.apache.org/`
- **MariaDB**: `https://mariadb.org/`
- **Datacharmer test database**: `https://github.com/datacharmer/test_db`
- **PHP**: `https://php.net/`
- **Adminer**: `https://www.adminer.org`
- **NGINX**: `https://nginx.org`

# 5

# Deploying WordPress

In the previous chapter, we built a playbook that installs and configures a basic **LAMP stack**. In this chapter, we will be building on top of the techniques we used there to create a playbook that installs a **LEMP stack**, which, as you might recall, replaces Apache with NGINX and then installs WordPress.

Once we finish this chapter, you should be able to do the following:

- Prepare our initial playbook
- Download and install the WordPress CLI
- Install and configure WordPress
- Log in to your WordPress installation

The chapter covers the following topics:

- Preinstallation tasks
- The `stack_install` role
- The `stack_config` role
- The `wordpress` role
- Running the WordPress playbook

Before we start, we should quickly cover what WordPress is; you have likely visited a website powered by WordPress at some point in the last 24 hours.

It is an open-source **content management system (CMS)** powered by PHP and MySQL and used by around 810 million websites, which is around 43% of all the websites on the internet today, according to the statistics published by Colorlib in August of 2023.

# Technical requirements

Like in *Chapter 4, Deploying a LAMP Stack*, we will use the local **Multipass virtual machine** we have been using throughout the title. Again, additional packages will be downloaded when launching the virtual machine and deploying WordPress.

You can find a complete copy of the playbook in the repository accompanying this title at `https://github.com/PacktPublishing/Learn-Ansible-Second-Edition/tree/main/Chapter05`.

# Preinstallation tasks

As mentioned in *Chapter 4, Deploying a LAMP Stack*, a LEMP stack is composed of the following elements:

- **Linux**: In our case, this will be the Ubuntu Multipass virtual machine
- **NGINX**: If you remember, it is pronounced as *engine-x*, which means there is an *E* in *LEMP* and not an *N* (which would also make it impossible to pronounce as an abbreviation)
- **MariaDB**: As we have already seen, this will be the database component
- **PHP**: We will be using PHP 8 again for this

Before we install WordPress, we need to install and configure these components. Also, as this playbook will eventually be executed against publicly available cloud servers, we must consider some best practices around our NGINX configuration.

However, before we start looking at the playbook, let's start things off by getting the initial structure of the playbook set up:

```
$ mkdir Chapter05 Chapter05/group_vars Chapter05/roles
$ touch Chapter05/group_vars/common.yml Chapter05/hosts Chapter05/
site.yml
$ cd Chapter05
```

This gives us our basic layout. Next, we must copy the `cloud-init.yaml`, `example_key`, `example_key.pub`, and `hosts.example` files from the previous chapters, so when it is time to run the playbook, we have everything we need to launch the virtual machine using Multipass.

Now that we have the basics configured, we can make a start by writing the playbook to deploy and configure our initial software stack.

# The stack_install role

We are going to start by creating a role called `stack_install` using `ansible-galaxy role init`:

```
$ ansible-galaxy role init roles/stack_install
```

This will install our initial software stack. Once installed, we hand it over to a second role, which will then configure the software stack before a third role starts the WordPress installation.

So, what packages do we need? WordPress has the following requirements:

- PHP 7.4 or greater
- MySQL 5.7 or greater OR MariaDB 10.4 or greater
- Nginx or Apache with the `mod_rewrite` module
- HTTPS support

We know from the previous chapter that the versions of PHP and MariaDB we are installing meet this requirement, leaving just NGINX, which we can download and install from the principal NGINX repository to get the latest and greatest version.

## Enabling the NGINX repository

Before we look at the tasks and variables that we will need to enable the mainline NGINX repository, let's start off the `roles/stack_install/tasks/main.yml` file with a task that updates the operating system and the cache of available packages:

```
- name: "Update apt-cache and upgrade packages"
  ansible.builtin.apt:
    name: "*"
    state: "latest"
    update_cache: true
```

The remainder of the tasks we will be defining enable the repository before we finally install the packages.

Moving onto the `roles/stack_install/default/main.yml` file, we need to set some variables containing information on the repository, which we will add alongside the default Ubuntu ones.

These variables start with one that contains the URL of the signing key for the repository that will be enabled:

```
repo_keys_url:
  - "http://nginx.org/keys/nginx_signing.key"
```

We will then add the following repository URLs:

```
repo_packages:
  - "deb http://nginx.org/packages/mainline/ubuntu/ {{ ansible_
distribution_release }} nginx"
  - "deb-src http://nginx.org/packages/mainline/ubuntu/ {{ ansible_
distribution_release }} nginx"
```

You may have noticed that we are using the ansible_distribution_release fact to dynamically run into the URL to put the correct version number of the Ubuntu distribution.

Now, back to the roles/stack_install/tasks/main.yml file and the two tasks that call these variables – these will look like the following, starting with the addition of the signing key:

```
- name: "Add the apt keys from a URL"
  ansible.builtin.apt_key:
    url: "{{ item }}"
    state: "present"
  with_items: "{{ repo_keys_url }}"
```

As you can see, we are using with_items, so, if you need to, you could define more than one URL and add additional signing keys.

This approach is carried forward to the next task, where we are adding more than one repository:

```
- name: "Install the repo packages"
  ansible.builtin.apt_repository:
    repo: "{{ item }}"
    state: "present"
    update_cache: true
  with_items: "{{ repo_packages }}"
```

The final task in the roles/stack_install/tasks/main.yml file is the one that installs all of the packages:

```
- name: "Update cache and install the stack packages"
  ansible.builtin.apt:
    state: "present"
    update_cache: true
    pkg: "{{ system_packages + extra_packages + stack_packages }}"
```

You will notice that rather than defining the packages in a single variable, I have split them into three, and we are combining them by using + when calling the variables.

So, what do these three variables contain, and why don't we define them as a single variable?

Back to the `roles/stack_install/default/main.yml` file, you can see that `system_packages` is defined as the following:

```
system_packages:
    - "software-properties-common"
    - "python3-pymysql"
    - "acl"
```

Following that, the `extra_packages` variable contains the following package list:

```
extra_packages:
    - "vim"
    - "git"
    - "unzip"
```

Finally, we have the list of packages that make up the bulk of our software stack:

```
stack_packages:
    - "nginx"
    - "mariadb-server"
    - "mariadb-client"
    - "php-cli"
    - "php-curl"
    - "php-fpm"
    - "php-gd"
    - "php-intl"
    - "php-mbstring"
    - "php-mysql"
    - "php-soap"
    - "php-xml"
    - "php-xmlrpc"
    - "php-zip"
```

As we are defining three variables for the packages, it means that we can, if required, overwrite them elsewhere in our playbook.

Let us, for example, assume that we need to install the Amazon Web Services command-line tool on the virtual machine.

This would allow us to push data, such as images, to an Amazon S3 bucket or clear a cache on a CloudFront content delivery network endpoint.

Rather than overriding a long list of packages from a single variable, we could take the `extra_packages` variable, add it to `group_vars/common.yml`, and append it to the end of the list of packages so that it will now look like the following:

```
extra_packages:
  - "vim"
  - "git"
  - "unzip"
  - "awscli"
```

As you can see, this is a lot more efficient than repeating all the packages we want to install.

Another advantage of using + to combine everything is that we only need to call a single `ansible.builtin.apt` task to install everything we need for the following role, which we will dive into now.

## The stack_config role

Now that we have our base software stack installed we need to configure it, let's start by creating the role by running the following command:

```
$ ansible-galaxy role init roles/stack_config
```

This gives us the basic file structure needed for the `stack_config` role. With that in place, we can now look at configuring the role itself – in this role, we will need to do the following:

- Add a system user for our WordPress installation to run under
- Configure NGINX as per the best practices on the WordPress documentation
- Configure PHP-FPM to run as the WordPress user we created earlier

As we need a user for WordPress to run under, we should make a start there.

### WordPress system user

The defaults for the WordPress system user, which should be placed in `roles/stackconfig/defaults/main.yml`, are as follows:

```
wordpress_system:
  user: "wordpress"
  group: "php-fpm"
  comment: "wordpress system user"
  home: "/var/www/wordpress"
  state: "present"
```

We refer to this as the system user, as we will create a user in WordPress itself later in the chapter. This user's details will also be defined in Ansible, so we do not want to get the two different users confused.

The two tasks that use these variables, found in `roles/stack_config/tasks/main.yml`, should look like this:

```
- name: "add the wordpress group"
  ansible.builtin.group:
    name: "{{ wordpress_system.group }}"
    state: "{{ wordpress_system.state }}"
```

The preceding task ensures that the group is present, and the next task adds an operating system-level user, which is added to the group that has just been created:

```
- name: "Add the wordpress user"
  ansible.builtin.user:
    name: "{{ wordpress_system.user }}"
    group: "{{ wordpress_system.group }}"
    comment: "{{ wordpress_system.comment }}"
    home: "{{ wordpress_system.home }}"
    state: "{{ wordpress_system.state }}"
```

As you can see, we are not adding a key to the user this time as we don't want to log in to the user account to start manipulating files and other actions. This should all be done within WordPress itself or by using Ansible.

## NGINX configuration

We are going to be using several template files for our NGINX configuration. The first template is called `roles/stack_config/templates/nginx-nginx.conf.j2`, and it will replace the main NGINX configuration deployed by the package installation:

```
# {{ ansible_managed }}
user  nginx;
worker_processes  {{ ansible_processor_count }};
error_log  /var/log/nginx/error.log warn;
pid        /var/run/nginx.pid;
events {
    worker_connections  1024;
}
http {
    include       /etc/nginx/mime.types;
    default_type  application/octet-stream;
    log_format  main  '$remote_addr - $remote_user [$time_local]
"$request" '
```

```
                        '$status $body_bytes_sent "$http_referer" '
                        '"$http_user_agent" "$http_x_forwarded_for"';
    access_log  /var/log/nginx/access.log  main;
    sendfile          on;
    keepalive_timeout  65;
    client_max_body_size 20m;
    include /etc/nginx/conf.d/*.conf;
}
```

The file's content is the same as the file that will be replaced, except that we are updating `worker_processes` to use the number of processors detected by Ansible when the setup module runs rather than a hardcoded value.

The task to deploy the configuration file is as you would expect, and it should be placed in `roles/stack_config/tasks/main.yml`:

```
- name: "Copy the nginx.conf to /etc/nginx/"
  ansible.builtin.template:
    src: nginx-nginx.conf.j2
    dest: /etc/nginx/nginx.conf
    mode: "0644"
  notify: "Restart nginx"
```

As you can see, we are notifying the restart `nginx` handler, which is stored in the `roles/stack_config/handlers/main.yml` file:

```
- name: "Restart nginx"
  ansible.builtin.service:
    name: nginx
    state: restarted
    enabled: true
```

Next, we have the default site template, `roles/stack_config/templates/nginx-confd-default.conf.j2`:

```
# {{ ansible_managed }}
upstream {{ php.upstream }} {
        server {{ php.ip }}:{{ php.port }};
}
server {
    listen          80;
    server_name  {{ ansible_nodename }};
    root            {{ wordpress_system.home }};
    index          index.php index.html index.htm;
```

```
    include global/restrictions.conf;
    include global/wordpress_shared.conf;
}
```

To help identify where the template files will be placed on the target host, I am naming them so that the full path is in the filename. In this case, the filename is `nginx-confd-default.conf.j2`, and it will be deployed to `/etc/nginx/conf.d/default.conf`; the task to do this follows:

```
- name: "Copy the default.conf to /etc/nginx/conf.d/"
  ansible.builtin.template:
    src: nginx-confd-default.conf.j2
    dest: /etc/nginx/conf.d/default.conf
    mode: "0644"
  notify: "Restart nginx"
```

The following two files we are deploying are going into a folder that doesn't exist. So, we first need to create the destination folder. To do this, we need to add the following to `roles/stack_config/tasks/main.yml`:

```
- name: "Create the global directory in /etc/nginx/"
  ansible.builtin.file:
    dest: /etc/nginx/global/
    state: directory
    mode: "0644"
```

As we are not making any replacements in the `nginx-global-restrictions.conf` file, we are using the `ansible.builtin.copy` module rather than `ansible.builtin.template` here; the file is stored in `roles/stack_config/files/` and the task that copies it is as follows:

```
- name: "Copy the restrictions.conf to /etc/nginx/global/"
  ansible.builtin.copy:
    src: nginx-global-restrictions.conf
    dest: /etc/nginx/global/restrictions.conf
    mode: "0644"
  notify: "Restart nginx"
```

This file has some sensible defaults in it, such as denying access to files that are included as part of the WordPress installation:

```
location ~* /(wp-config.php|readme.html|license.txt|nginx.conf) {
    deny all;
}
```

Another import inclusion is adding a configuration to deny access to .php files within /wp-content/ and its sub-folders:

```
location ~* ^/wp-content/.*.(php|phps)$ {
    deny all;
}
```

There are several other configurations in the nginx-global-restrictions.conf file; see the repository, which accompanies the book, for the complete configuration, as there are too many snippets for us to go into here.

The same can be said for the next and final block of the NGINX configuration; review the repository for more information on the configuration deployed by the following task:

```
- name: "Copy the wordpress_shared.conf to /etc/nginx/global/"
  ansible.builtin.template:
    src: nginx-global-wordpress_shared.conf.j2
    dest: /etc/nginx/global/wordpress_shared.conf
    mode: "0644"
  notify: "Restart nginx"
```

When we reviewed the default site template, roles/stack_config/templates/nginx-confd-default.conf.j2, you may have noticed the use of a few variables we haven't yet defined; they were php.ip and php.port.

As you may have already guessed from the variable labeling, these have to do with the configuration of PHP, so, let us look at configuring the PHP and PHP-FPM part of our deployment.

### PHP and PHP-FPM configuration

As we saw in the previous section, there are a few variables defined for PHP in roles/stack_config/defaults/main.yml, and these are as follows:

```
php:
  ip: "127.0.0.1"
  port: "9000"
  upstream: "php"
  ini:
    - { regexp: "^;date.timezone =", replace: "date.timezone = Europe/
London" }
    - { regexp: "^expose_php = On", replace: "expose_php = Off" }
    - {
        regexp: "^upload_max_filesize = 2M",
        replace: "upload_max_filesize = 20M",
      }
```

We then have some variables that define some information on the paths for the various files and service names:

```
php_fpm_path: "/etc/php/8.1/fpm/pool.d/www.conf"
php_ini_path: "/etc/php/8.1/fpm/php.ini"
php_service_name: "php8.1-fpm"
```

The first configuration of the two tasks we will be running deploys the PHP-FPM configuration; this is, what the template, which can be found at roles/stack_config/templates/php-fpmd-www.conf.j2, looks like:

```
; {{ ansible_managed }}
[{{ wordpress_system.user }}]
user = {{ wordpress_system.user }}
group = {{ wordpress_system.group }}
listen = {{ php.ip }}:{{ php.port }}
listen.allowed_clients = {{ php.ip }}
pm = dynamic
pm.max_children = 50
pm.start_servers = 5
pm.min_spare_servers = 5
pm.max_spare_servers = 35
php_admin_value[error_log] = /var/log/php-fpm/{{ wordpress_system.user
}}-error.log
php_admin_flag[log_errors] = on
php_value[session.save_handler] = files
php_value[session.save_path]    = /var/lib/php/fpm/session
php_value[soap.wsdl_cache_dir]  = /var/lib/php/fpm/wsdlcache
```

As you can see, we have a few replacements in this file. Starting at the top between the square brackets, we are defining the PHP-FPM pool name and using the content of the wordpress_system.user variable for this.

Next, we have the user and group we want our pool to run under; here, we use wordpress_system.user and wordpress_system.group.

Finally, we are setting the IP address and port we want our PHP-FPM pool to listen on by using the php.ip and php.port variables.

The task in roles/stack_config/tasks/main.yml to deploy the template looks as follows:

```
- name: "Copy the www.conf to /etc/php-fpm.d/"
  ansible.builtin.template:
    src: php-fpmd-www.conf.j2
    dest: "{{ php_fpm_path }}"
```

```
    mode: "0644"
  notify: "Restart php-fpm"
```

The handler to restart PHP-FPM in `roles/stack_config/handlers/main.yml` is very similar to the ones we have already been defining throughout the book:

```
- name: "Restart php-fpm"
  ansible.builtin.service:
    name: "{{ php_service_name }}"
    state: restarted
    enabled: true
```

The next task in `roles/stack_config/tasks/main.yml` uses the `ansible.builtin.lineinfile` module:

```
- name: "Configure php.ini settings"
  ansible.builtin.lineinfile:
    dest: "{{ php_ini_path }}"
    regexp: "{{ item.regexp }}"
    line: "{{ item.replace }}"
    backup: "true"
    backrefs: "true"
  with_items: "{{ php.ini }}"
  notify: "Restart php-fpm"
```

We are taking the `php.ini` file and looping through it by looking for the values defined by the `regexp` key. Once we find the value, we replace it with the content of the replace key. If there are changes to the file, we are making a backup first, just in case.

Also, we are using `backrefs` to ensure that if there is no matching `regex` in the file, then it will be left unchanged; if we didn't use them, the `restart php-fpm` handler would be called every time the playbook runs, and we do not want PHP-FPM to be restarted if there is no reason for it be.

### Starting NGINX and PHP-FPM

Now that we have NGINX and PHP-FPM installed and configured, we need to start the two services rather than wait until the end of the playbook run.

If we don't do this now, our upcoming role to install WordPress will fail. The first of the two tasks in `roles/stackconfig/tasks/main.yml` looks like the following:

```
- name: "Start php-fpm"
  ansible.builtin.service:
    name: "{{ php_service_name }}"
    state: "started"
```

The second task looks pretty much the same:

```
- name: "Start nginx"
  ansible.builtin.service:
    name: "nginx"
    state: "started"
```

If you look at the two tasks, they are the same as the two handlers we have already defined.

However, if you look closer, you will notice that while we are using the ansible.builtin. service module, we are only setting the state setting to started rather than restarted, and we are missing the configuration for enabled, which sets the service to start on boot.

Another thing you may have noticed is the use of the php_service_name variable; to explain why we are using this, you will need to wait until *Chapter 6, Targeting Multiple Distributions*.

The final component of our software stack that we need to configure is MariaDB, so let us review that before we move on to the WordPress installation and configuration.

### MariaDB configuration

The MariaDB configuration will closely match its configuration in *Chapter 4, Deploying a LAMP Stack*, minus a few steps, so I will not go into too much detail here.

The default variables for this part of the role in roles/stack_config/defaults/main. yml are as follows:

```
mariadb:
  bind: "127.0.0.1"
  server_config: "/etc/my.cnf.d/mariadb-server.cnf"
  username: "root"
  password: "Pa55W0rd123"
  hosts:
    - "127.0.0.1"
    - "::1"
    - "{{ ansible_nodename }}"
    - "localhost"
```

As you can see, we are now using a nested variable and have removed the host wildcard, which we had previously defined as % in *Chapter 4, Deploying a LAMP Stack*.

Our first task is to start MariaDB so that we can interact with it:

```
- name: "Start mariadb"
  ansible.builtin.service:
    name: "mariadb"
```

```
      state: "started"
      enabled: true
```

Check for the presence of the ~/.my.cnf file:

```
- name: "Check to see if the ~/.my.cnf file exists"
  ansible.builtin.stat:
    path: "~/.my.cnf"
  register: mycnf
```

Set a password:

```
- name: "Change mysql root password if we need to"
  community.mysql.mysql_user:
    name: "{{ mariadb.username }}"
    host: "{{ item }}"
    password: "{{ mariadb.password }}"
    check_implicit_admin: "true"
    priv: "*.*:ALL,GRANT"
    login_user: "{{ mariadb.username }}"
    login_unix_socket: /var/run/mysqld/mysqld.sock
  with_items: "{{ mariadb.hosts }}"
  when: not mycnf.stat.exists
```

Create the ~/my.cnf file:

```
- name: "Set up .my.cnf file"
  ansible.builtin.template:
    src: "my.cnf.j2"
    dest: "~/.my.cnf"
    mode: "0644"
```

Then, remove the anonymous user:

```
- name: "Delete anonymous MySQL user"
  community.mysql.mysql_user:
    user: ""
    host: "{{ item }}"
    state: "absent"
  with_items: "{{ mariadb.hosts }}"
```

Now, we have come to our final task, which is to remove the test database:

```
- name: "Remove the MySQL test database"
  community.mysql.mysql_db:
```

```
    db: "test"
    state: "absent"
```

Now, with everything we need to install and run WordPress configured, we can start on WordPress itself.

# The wordpress role

Now that we have completed the roles that prepare our target virtual machine, we can proceed with the actual WordPress installation; this will be split into a few different parts, starting with downloading `wp-cli` and setting up the database.

Before we progress, we should create the role:

```
$ ansible-galaxy role init roles/wordpress
```

Now that we have the empty role files, we can start populating the tasks and variables in the files.

## Some facts

Before installing WordPress, we must set some facts using the `ansible.builtin.set_fact` module. The following task, the first in the `roles/wordpress/tasks/main.yml` file, sets two variables using the information gathered when Ansible first connects to the hosts:

```
- name: "Set a fact for the wordpress domain"
  ansible.builtin.set_fact:
    wordpress_domain: "{{ ansible_ssh_host }}"
    os_family: "{{ ansible_distribution }} {{ ansible_distribution_
  version }}"
```

We will use these two variables when we install WordPress using the WordPress CLI, which we will be downloading and installing next.

## WordPress CLI installation

WordPress CLI (`wp-cli`) is a command-line tool used to administer your WordPress installation; we will be using it throughout the role, so, the first thing our role should do is download it. To do this, we need to download the following variables in `roles/wordpress/defaults/main.yml`:

```
wp_cli:
  download: "https://raw.githubusercontent.com/wp-cli/builds/gh-pages/
phar/wp-cli.phar"
  path: "/usr/local/bin/wp"
```

Moving back to the `roles/wordpress/tasks/main.yml` file, we use these two variables in the following task, which downloads `wp-cli` and places it on our host:

```
- name: "Download wp-cli"
  ansible.builtin.get_url:
    url: "{{ wp_cli.download }}"
    dest: "{{ wp_cli.path }}"
    mode: "0755"
```

Now, we have `wp-cli` on our host with the correct execute permissions.

Before we start to use `wp-cli`, we have one more bit of preparation work to do: create the database and user, which we will use with our WordPress installation.

## Creating the WordPress database

The next part of the role creates the database our WordPress installation will use; as per the other tasks in this chapter, it uses a nested variable, which can be found in `roles/wordpress/defaults/main.yml`:

```
wp_database:
  name: "wordpress"
  username: "wordpress"
  password: "WO4DPr3S5"
```

The tasks in `roles/wordpress/tasks/main.yml` to create the database are as follows:

```
- name: "Create the wordpress database"
  community.mysql.mysql_db:
    db: "{{ wp_database.name }}"
    state: "present"
```

Now that the database has been created, we can add the user:

```
- name: "Create the user for the wordpress database"
  community.mysql.mysql_user:
    name: "{{ wp_database.username }}"
    password: "{{ wp_database.password }}"
    priv: "{{ wp_database.name }}.*:ALL"
    state: "present"
  with_items: "{{ mariadb.hosts }}"
```

Notice how we are using the `mariadb.hosts` variable from the previous role. Now that we have the database created, we can start downloading and installing WordPress.

## Downloading, configuring, and installing WordPress

Now that we have everything in place to install WordPress, we can make a start, first by setting some default variables in `roles/wordpress/defaults/main.yml`:

```
wordpress:
  domain: "http://{{ wordpress_domain }}/"
  title: "WordPress installed by Ansible on {{ os_family }}"
  username: "ansible"
  password: "password"
  email: "test@example.com"
  plugins:
    - "jetpack"
    - "wp-super-cache"
    - "wordpress-seo"
    - "wordfence"
    - "nginx-helper"
```

Now that we have our variables, we can start our download if we need to. To find out whether we need to download WordPress, we should check for the presence of an existing WordPress installation. The task to do this in `roles/wordpress/tasks/main.yml` looks like the following:

```
- name: "Are the wordpress files already there?"
  ansible.builtin.stat:
    path: "{{ wordpress_system.home }}/index.php"
  register: wp_installed
```

As you can see, the first task uses the `ansible.builtin.stat` module to check for an `index.php` file in our system user's home directory, which in our case is also the webroot.

If this is the first time that the playbook is being run against the host, then we will need to download WordPress:

```
- name: "Download wordpresss"
  ansible.builtin.command: "{{ wp_cli.path }} core download"
  args:
    chdir: "{{ wordpress_system.home }}"
  become_user: "{{ wordpress_system.user }}"
  become: true
  when: not wp_installed.stat.exists
```

This task uses the `ansible.builtin.shell` module to issue the following command:

```
$ su wordpress -
$ cd /var/www/wordpress
$ /usr/local/bin/wp core download
```

There are a few arguments we should work through before moving on to the next task, which are the following:

- args and chdir: You can pass additional arguments to the ansible.builtin.shell module using args. Here, we are passing chdir, which instructs Ansible to change to the directory we specify before running the shell command we provide.

- become_user: This is the user we want to run the command as. The command will run as the root user if we do not use this flag.

- become: This instructs Ansible to execute the task as the defined user.

The next task in the playbook sets the correct permissions on the user's home directory:

```
- name: "Set the correct permissions on the homedir"
  ansible.builtin.file:
    path: "{{ wordpress_system.home }}"
    mode: "0755"
  when: not wp_installed.stat.exists
```

Now that WordPress is downloaded, we can start the installation. First, we need to check whether this has already been done:

```
- name: "Is wordpress already configured?"
  ansible.builtin.stat:
    path: "{{ wordpress_system.home }}/wp-config.php"
  register: wp_configured
```

If there is no wp-config.php file, then the following task will be executed:

```
- name: "Sort the basic wordpress configuration"
  ansible.builtin.command: "{{ wp_cli.path }} core config --dbhost={{
mariadb.bind }} --dbname={{ wp_database.name }} --dbuser={{ wp_
database.username }} --dbpass={{ wp_database.password }}"
  args:
    chdir: "{{ wordpress_system.home }}"
  become_user: "{{ wordpress_system.user }}"
  become: true
  when: not wp_configured.stat.exists
```

This is like you logging in and running the following:

```
$ su wordpress -
$ cd /var/www/wordpress
$ /usr/local/bin/wp core config \
--dbhost=127.0.0.1\
```

```
--dbname=wordpress\
--dbuser=wordpress \
--dbpass=W04DPr3S5
```

As you can see, we are using Ansible to execute commands as if we had a local terminal open.

Now that we have our wp-config.php file created, with the database credentials in place, we can install WordPress.

First, we need to check whether WordPress has already been installed:

```
- name: "Do we need to install wordpress?"
  ansible.builtin.command: "{{ wp_cli.path }} core is-installed"
  args:
    chdir: "{{ wordpress_system.home }}"
  become_user: "{{ wordpress_system.user }}"
  become: true
  ignore_errors: true
  register: wp_installed
```

As you can see from the presence of the ignore_errors option, if WordPress is not installed, this command will give us an error. We are then using this to our advantage when registering the results, as you can see from the following task:

```
- name: "Install wordpress if needed"
  ansible.builtin.command: "{{ wp_cli.path }} core install --url='{{
  wordpress.domain }}' --title='{{ wordpress.title }}' --admin_user={{
  wordpress.username }} --admin_password={{ wordpress.password }}
  --admin_email={{ wordpress.email }}"
  args:
    chdir: "{{ wordpress_system.home }}"
  become_user: "{{ wordpress_system.user }}"
  become: true
  when: wp_installed.rc == 1
```

This task is only executed if the previous task returns an error, which is what happens if WordPress is not installed.

Now that our primary WordPress site is installed, we can continue installing the plugins.

## WordPress plugins installation

The final part of our WordPress installation is to download and install the plugins we defined in the wordpress.plugins variable.

As per previous tasks, we will build a little logic into the tasks. First, we run the following task to see whether all the plugins are already installed:

```
- name: "Do we need to install the plugins?"
  ansible.builtin.command: "{{ wp_cli.path }} plugin is-installed {{
item }}"
  args:
    chdir: "{{ wordpress_system.home }}"
  become_user: "{{ wordpress_system.user }}"
  become: true
  with_items: "{{ wordpress.plugins }}"
  ignore_errors: true
  register: wp_plugin_installed
```

If the plugins are not installed, this task should fail, so we have `ignore_errors` in there.

As you can see, we are registering the results of the entire task, because, if you remember, we are installing several plugins, such as `wp_plugin_installed`.

The next two tasks take the results of `wp_plugin_installed` and use the `ansible.builtin.set_fact` module to set a fact depending on the results:

```
- name: "Set a fact if we don't need to install the plugins"
  ansible.builtin.set_fact:
    wp_plugin_installed_skip: true
  when: wp_plugin_installed.failed is undefined
```

The preceding task is set if we don't need to install any of the plugins, and the following one is used if we need to install at least one of the plugins:

```
- name: "Set a fact if we need to install the plugins"
  ansible.builtin.set_fact:
    wp_plugin_installed_skip: false
  when: wp_plugin_installed.failed is defined
```

As you can see, we are setting `wp_plugin_installed_skip` to be `true` or `false`: if the fact is set to `false`, then the next task will loop through installing the plugins:

```
- name: "Install the plugins if we need to or ignore if not"
  ansible.builtin.command: "{{ wp_cli.path }} plugin install {{ item
}} --activate"
  args:
    chdir: "{{ wordpress_system.home }}"
```

```
become_user: "{{ wordpress_system.user }}"
become: true
with_items: "{{ wordpress.plugins }}"
when: not wp_plugin_installed_skip
```

Now that we have the plugins' tasks defined, we can have a go at running our playbook.

## Running the WordPress playbook

To run the playbook and install WordPress, we need to finish walking through the files; site.yml should look as follows:

```
---


- name: "Install and configure WordPress and supporting software"
  hosts: "ansible_hosts"
  gather_facts: true
  become: true
  become_method: "ansible.builtin.sudo"

  vars_files:
    - "group_vars/common.yml"

  roles:
    - "stack_install"
    - "stack_config"
    - "wordpress"
```

With that out of the way, we can run the playbook.

> **Note**
>
> The Chapter05 folder in the GitHub repository accompanying this title contains the example hosts file and keys for launching a local virtual machine using Multipass. If you are following along, please refer to the instructions in *Chapter 1, Installing and Running Ansible*; these detail how to launch the virtual machine and prepare your hosts file.

As we know, to run the playbook, we need to issue the following command once our Multipass virtual machine is up and running:

```
$ ansible-playbook -i hosts site.yml
```

Let's cover some of the highlights rather than go through the whole output here, starting with adding the NGINX repository:

```
TASK [roles/stack_install : add the apt keys from a URL] **
changed: [ansiblevm] => (item=http://nginx.org/keys/nginx_signing.key)
TASK [roles/stack_install : install the repo packages] ****
changed: [ansiblevm] => (item=deb http://nginx.org/packages/mainline/
ubuntu/ jammy nginx)
changed: [ansiblevm] => (item=deb-src http://nginx.org/packages/
mainline/ubuntu/ jammy nginx)
```

As you can see, the name of the Ubuntu release is added – in the example, this is jammy.

When making changes to the php.ini file, only two of the three changes we defined need to be applied, as expose_php is already set to Off:

```
TASK [roles/stack_config : configure php.ini] *************
changed: [ansiblevm] => (item={'regexp': '^;date.timezone =',
'replace': 'date.timezone = Europe/London'})
ok: [ansiblevm] => (item={'regexp': '^expose_php = On', 'replace':
'expose_php = Off'})
changed: [ansiblevm] => (item={'regexp': '^upload_max_filesize = 2M',
'replace': 'upload_max_filesize = 20M'})
```

Remember that we set the ignore_errors flag for some of the checks when it came to installing and configuring WordPress; this is why:

```
TASK [roles/wordpress : do we need to install wordpress?] *
fatal: [ansiblevm]: FAILED! => {"changed": true, "cmd": "/usr/local/
bin/wp core is-installed", "delta": "0:00:00.142910", "end": "2023-09-
17 12:28:16.500304", "msg": "non-zero return code", "rc": 1, "start":
"2023-09-17 12:28:16.357394", "stderr": "PHP Warning:  Undefined
array key \"HTTP_HOST\" in /var/www/wordpress/wp-includes/functions.
php on line 6135\nWarning: Undefined array key \"HTTP_HOST\" in /
var/www/wordpress/wp-includes/functions.php on line 6135\nPHP
Warning:  Undefined array key \"HTTP_HOST\" in /var/www/wordpress/
wp-includes/functions.php on line 6135\nWarning: Undefined array key
\"HTTP_HOST\" in /var/www/wordpress/wp-includes/functions.php on line
6135", "stderr_lines": ["PHP Warning:  Undefined array key \"HTTP_
HOST\" in /var/www/wordpress/wp-includes/functions.php on line 6135",
"Warning: Undefined array key \"HTTP_HOST\" in /var/www/wordpress/
wp-includes/functions.php on line 6135", "PHP Warning:  Undefined
array key \"HTTP_HOST\" in /var/www/wordpress/wp-includes/functions.
php on line 6135", "Warning: Undefined array key \"HTTP_HOST\" in /
var/www/wordpress/wp-includes/functions.php on line 6135"], "stdout":
"", "stdout_lines": []}
...ignoring
TASK [roles/wordpress : install wordpress if needed] ******
changed: [ansiblevm]
```

As you can see, an error was ignored, and the task to install WordPress was triggered. The same thing happened for the plugins:

```
TASK [roles/wordpress : set a fact if we don't need to install the
plugins] **************************************
skipping: [ansiblevm]
TASK [roles/wordpress : set a fact if we need to install the plugins]
***********************************************
ok: [ansiblevm]
```

On first execution, the recap looked something like the following:

```
PLAY RECAP ****************************************************
ansiblevm                    : ok=39    changed=28    unreacha-
ble=0     failed=0    skipped=1    rescued=0    ignored=2
```

Rerunning the playbook immediately after shows how the logic we have added throughout the task execution kicks in, which results in a lot of the later tasks being skipped entirely:

```
TASK [roles/wordpress : are the wordpress files already there?] ******
*************************************************
ok: [ansiblevm]
TASK [roles/wordpress : download wordpresss] **************
skipping: [ansiblevm]
```

Note that, this time, the check for the plugins doesn't result in an error:

```
TASK [roles/wordpress : do we need to install the plugins?]
changed: [ansiblevm] => (item=jetpack)
changed: [ansiblevm] => (item=wp-super-cache)
changed: [ansiblevm] => (item=wordpress-seo)
changed: [ansiblevm] => (item=wordfence)
changed: [ansiblevm] => (item=nginx-helper)

TASK [roles/wordpress : set a fact if we don't need to install the
plugins] **************************************
ok: [ansiblevm]
TASK [roles/wordpress : set a fact if we need to install the plugins]
***********************************************
skipping: [ansiblevm]
TASK [roles/wordpress : install the plugins if we need to or ignore if
not] ******************************************
skipping: [ansiblevm] => (item=jetpack)
skipping: [ansiblevm] => (item=wp-super-cache)
skipping: [ansiblevm] => (item=wordpress-seo)
skipping: [ansiblevm] => (item=wordfence)
skipping: [ansiblevm] => (item=nginx-helper)
```

Now that WordPress is installed, we should be able to access it in a browser by going to the host you have defined in your `hosts` file, in my case, `http://192.168.64.26.nip.io/`; yours will be different.

You will see the default WordPress site:

Figure 5.1 – Our freshly installed WordPress website

As you can see, the site's description in the top left reads **WordPress installed by Ansible on Ubuntu 22.04**, which is what we set when installing WordPress.

Also, if you go to the WordPress admin area by appending `/wp-admin/` to the end of your URL, for example, `http://192.168.64.26.nip.io/wp-admin/`, you should be able to log in to WordPress using the username and password we defined:

Figure 5.2 – The WordPress admin login page

Once logged in, you should see a few messages about the plugins we installed during the playbook run needing to be configured:

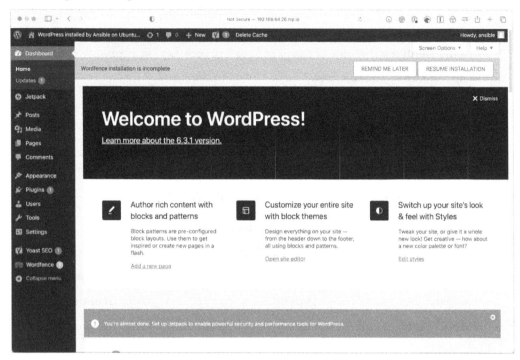

Figure 5.3 – Prompts when first logging into WordPress

Feel free to play with the WordPress installation and even, if you are so inclined, try and break it – if you needed to, you could delete and relaunch the Multipass virtual machine and quickly rerun the playbook to reinstall WordPress.

## Summary

In this chapter, we have reused many of the same principles we covered in the previous chapter and moved on to deploying a complete application. What is good about this is that the process is both repeatable and just a single command.

So far, we have been targeting an Ubuntu virtual machine. If we ran our playbook against a Red-Hat-based virtual machine, the playbook would give an error as commands and paths are different.

The next chapter will target multiple operating systems using the same playbook.

# Further reading

You can find out more information on the technologies we have covered in this chapter at the following links:

- **Colorlib WordPress statistics**: `https://colorlib.com/wp/wordpress-statistics`
- **NGINX**: `http://nginx.org/`
- **WordPress**: `https://wordpress.org/`
- **WP-CLI**: `http://wp-cli.org/`
- **WordPress on NGINX**: `https://wordpress.org/documentation/article/nginx/`

The project pages for the plugins we installed can be found at the following links:

- **Jetpack**: `https://en-gb.wordpress.org/plugins/jetpack/`
- **WP Super Cache**: `https://en-gb.wordpress.org/plugins/wp-super-cache/`
- **Yoast SEO**: `https://en-gb.wordpress.org/plugins/wordpress-seo/`
- **Wordfence**: `https://en-gb.wordpress.org/plugins/wordfence/`
- **NGINX Helper**: `https://wordpress.org/plugins/nginx-helper/`

# 6

# Targeting Multiple Distributions

So far, throughout the previous chapters, we have been targeting a single operating system, Ubuntu, when running on our playbooks.

This chapter will examine how to work with multiple Linux distributions within the same roles and playbooks.

We will take the WordPress playbook and the roles we created in *Chapter 5, Deploying WordPress*, and do the following:

- Discover what the difference is between our two target operating systems
- Look at and implement our WordPress roles, making them work on both target operating systems
- Discuss and apply best practices for targeting multiple distributions

The chapter covers the following topics:

- Debian and Red Hat
- Multi-distribution considerations
- Adapting the roles
- Running the playbook

## Technical requirements

Given that we will be launching two different operating systems, we will be changing the approach that we have taken in previous chapters and launching a pair of virtual machines in a cloud provider rather than two different virtual machines on our local machines.

The primary reason for this is that Multipass only really supports Ubuntu machines as it was created by Canonical, the creators and maintainers of Ubuntu, to give people a quick, easy, and consistent way to launch an Ubuntu virtual machine across multiple host platforms.

As we will be looking at automating cloud deployments in *Chapter 9, Moving to the Cloud*, we won't use Ansible to deploy the cloud resources for this chapter.

For this chapter, I would recommend using a cloud provider such as **DigitalOcean** (`https://www.digitalocean.com/`) or **Linode** (`http://www.linode.com/`), both of whom support the operating systems we will be covering in this chapter and whose virtual machine costs start at around 5 USD per month.

> **Important note**
>
> This chapter will not cover how to launch virtual machines; if you are following along, please review your chosen cloud provider's documentation. Additionally, for the full working code, please see the GitHub repository at `https://github.com/PacktPublishing/Learn-Ansible-Second-Edition`.

# Debian and Red Hat

This is where the world of Linux operating systems can get slightly confusing. Although we launch Ubuntu `22.04` and Rocky Linux `9` virtual machines to run our playbooks against, we will reference Debian and Red Hat within the playbook code.

Why is that? The reason behind this lies in the lineage of Linux distributions. Ubuntu is a descendant of the Debian operating system, inheriting its package management system and many other features. Similarly, Rocky Linux is a descendant of Red Hat, designed to be a downstream, bug-for-bug compatible release with **Red Hat Enterprise Linux** (**RHEL**).

So, when we mention Debian and Red Hat in our playbooks, we're referring to the fundamental bases from which our two operating systems, Ubuntu and Rocky Linux, have evolved.

In practical terms, the playbook code will often check the underlying distribution type to determine how to proceed with specific tasks. For example, the commands to install a software package on a Debian-based system such as Ubuntu might differ from those on a Red Hat-based system such as Rocky Linux.

Debian-based systems use the Debian package management system, with `dpkg` as the core utility, and often utilize either `apt` or `apt-get`, or in some cases all of them, for user-friendly interactions.

Red Hat-based systems employ the RPM package management system, using `rpm` as the core utility, often complemented by `yum` or its successor `dnf` for a more user-friendly interface for managing packages.

There are other differences, such as Debian and Red Hat-based systems that have different directory structures and configuration file locations, which can affect the system administration that we must consider in our playbook roles.

The biggest, at the time of writing, and most relevant difference between the two is licensing.

Debian is known for its strict adherence to free software principles. In contrast, Red Hat-based systems may incorporate more proprietary or closed-source software, especially in the case of Red Hat Enterprise Linux's commercial enterprise distribution of Red Hat.

This came to a head in June 2023 when Red Hat altered its terms, ceasing the public availability of RHEL's source code and restricting access solely to customers.

This move impacted downstream projects, relying on RHEL source code to create compatible distributions such as Rocky Linux. The change means that only customers bound by contracts preventing code sharing can access RHEL source code, aligning with the GPL license's terms, which mandates source code availability only for binary users, who are, essentially, the paying customers in this scenario.

At the time of writing, the fallout from this change is still being felt, and the dust is still settling, although it does seem like distributions such as Rocky Linux have found ways of being compliant; see the *Further reading* section at the end of this chapter for more information.

So, back to our playbook, by referencing either (or both) Debian and Red Hat in the code, we create more adaptable roles that can handle different Linux distributions and their derivatives consistently.

## Multi-distribution considerations

Looking at each of the Ansible built-in modules used in the three roles, `stack_install`, `stack_config`, and `wordpress`, we are using a few that will not work on our newly introduced Rocky Linux box.

Let's quickly work through each module and consider what we need to change or take into account when targeting two different distributions.

### The Stack Install role

This role uses the following built-in modules:

- `ansible.builtin.apt`
- `ansible.builtin.apt_key`
- `ansible.builtin.apt_repository`

We use these modules to update our operating system, add the NGINX mainline repository, and install all the packages we require for our WordPress installation.

As these modules all deal with package management, we won't be able to reuse any of these tasks, meaning that we will need to split the role into two parts: one that deals with Debian-based systems and the other for Red Hat systems.

Additionally, we won't be able to reuse the variables, as there are subtle differences in the package names between the two distributions.

This means that our best approach to this role is to use two different sets of tasks depending on the distribution Ansible is targeting. Luckily, there are built-in Ansible modules that make this approach simple. After reviewing the modules in the two remaining roles, we will cover these in the next section.

## The Stack Config role

This role is slightly different from the previous one in that we don't need to split the tasks into two here; most of the tasks will work across both our Linux distributions.

This means that the tasks which make use of the following modules won't need any changes:

- `ansible.builtin.group`: Creating a group is the same for both distributions
- `ansible.builtin.user`: Creating a user is the same for both distributions
- `ansible.builtin.template`: This only renders and copies files to the target hosts
- `ansible.builtin.file`: This only copies files to the target hosts
- `ansible.builtin.copy`: This only copies files on the target hosts
- `ansible.builtin.lineinfile`: This only searches for text and, if required, updates it within the files on the target hosts
- `ansible.builtin.service`: This is supported on both distributions
- `ansible.builtin.stat`: Only checks for the presence of a file on the host's file system
- `ansible.builtin.mysql_user`: As this interacts with the database service, it is distribution agnostic
- `ansible.builtin.mysql_db`: As with the previous task, it interacts with the database service

This list is mostly true; however, the file paths will change between the two distributions.

Still, as we already mentioned in *Chapter 5, Deploying WordPress*, when we looked at the variables for the Stack Config role, we are referencing files that contain the variables we want to load into the playbook run, so we will need to load in an additional set of variables for the distribution as well as the standard ones.

We will need to execute some additional tasks as part of adding the second distribution. Some Red Hat distributions come with a firewall enabled out of the box and SELinux enabled, so we will need to perform some Red Hat-only tasks at the end.

**SELinux**, or to give it its full name, **Security-Enhanced Linux**, is a security module of the Linux kernel that provides a mechanism for supporting access control security policies.

However, we can keep these tasks within the `main.yml` file rather than loading a different set of tasks by getting creative with the conditions when calling the tasks.

### The WordPress role

As the previous two roles have already installed and configured everything that we need to run our WordPress installation, this role is entirely distribution agnostic, and we don't need to make any changes to the tasks within the role. If you remember, in *Chapter 5, Deploying WordPress*, when we ran the command to configure WordPress, we set the following fact:

```
- name: "Set a fact for the wordpress domain"
  ansible.builtin.set_fact:
    wordpress_domain: "{{ ansible_ssh_host }}"
    os_family: "{{ ansible_distribution }} {{ ansible_distribution_
version }}"
```

This used the facts gathered by Ansible when first connecting to the host to figure out which distribution and version we were connecting to; we will expand on this logic as we dive deeper into the changes outlined in this section for the Stack Install and Config roles.

# Adapting the roles

So, how do we build the logic into our roles to execute only certain parts of them on different operating systems? As we know, the package names will be different. How do we define different sets of variables per operating system?

### Operating system family

We have looked at the `ansible.builtin.setup` module in *Chapter 1, Installing and Running Ansible*; this module gathers facts about our target hosts.

One of these facts is `ansible_os_family`; this tells us the type of operating system we are running.

To demonstrate this, I have launched two hosts, one running Ubuntu 22.04, and the second running Rocky Linux 9 as its operating system. I have created an inventory file which looks like the following:

```
RedHat ansible_host=178.79.178.78.nip.io
Debian ansible_host=176.58.114.60.nip.io
[ansible_hosts]
RedHat
Debian
[ansible_hosts:vars]
```

```
ansible_connection=ssh
ansible_user=root
ansible_private_key_file=~/.ssh/id_rsa
host_key_checking=False
```

> **Important note**
>
> The preceding inventory file is only for illustrative purposes; if you are following along, you will need to update it to consider your host IP addresses, user names, and private key file locations.

With the hosts up and running, we can target each one individually using the following commands:

```
$ ansible -i hosts RedHat -m ansible.builtin.setup | grep ansible_os_
family
$ ansible -i hosts Debian -m ansible.builtin.setup | grep ansible_os_
family
```

Running these two commands should show you something like the following terminal output:

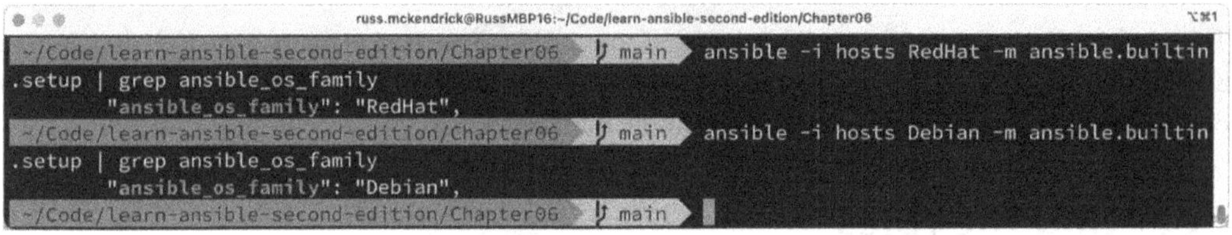

Figure 6.1 – Checking the values of ansible_os_family

As you can see, each of the two hosts correctly returns the operating system family.

We can take this one step further and update our commands to the following:

```
$ ansible -i hosts RedHat -m ansible.builtin.setup | grep ansible_
distribution
$ ansible -i hosts Debian -m ansible.builtin.setup | grep ansible_
distribution
```

This gives the following output:

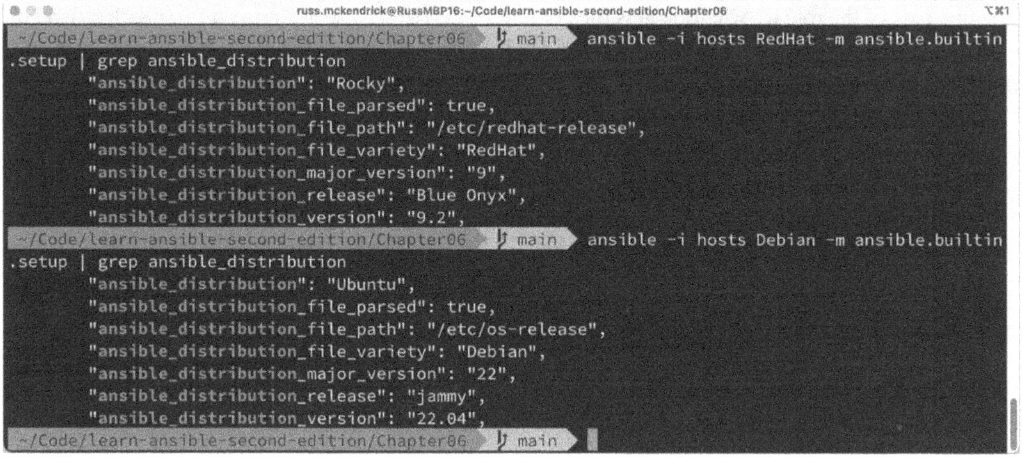

Figure 6.2 – Checking the values of ansible_distribution

As you can see, this gives much more detail on the operating system itself and not just the flavor of Linux; it is based on `RedHat` or `Debian`.

Finally, we run the following command:

```
$ ansible -i hosts ansible_hosts -m ansible.builtin.setup | grep
ansible_os_family
```

This will target both hosts within the same Ansible run and return a terminal output that should look like the following:

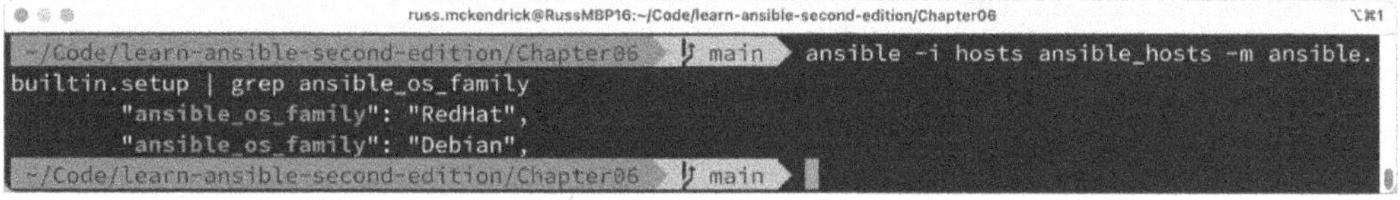

Figure 6.3 – Checking the values of ansible_distribution in a single run

Now that we can identify which operating system is in use on each host, we can start adapting the roles to consider the changes we discussed in the previous section of this chapter.

## The Stack Install role

The first part of the role we will look at is the content of roles/stack_install/tasks/main. yml. The previous version of the role contained all of the tasks to install the repos and packages for our Ubuntu server; all of those tasks should be moved to a file called roles/stack_install/ tasks/Debian.yml, and a new file called roles/stack_install/tasks/RedHat.yml should have been created; finally, we should update roles/stack_install/tasks/main. yml so that it has the following contents.

Here are the three task loads in the variables file for the operating system we are targeting:

```
- name: "Include the operating system specific variables"
  ansible.builtin.include_vars: "{{ ansible_os_family }}.yml"
```

As you can see, this uses the ansible.builtin.include_vars module to load variables from the variables path within the roles folder, which would be roles/stack_install/vars/.

Then, it loads a file called RedHat.yml or Debian.yml; these two file names are populated using the {{ ansible_os_family }} variable in the task, meaning that the variables relevant to the operating system being targeted are loaded.

If you look in the repository on GitHub, you will notice that, although being subtle, there are differences in the packages listed in the system_packages, extra_packages, and stack_packages package lists.

The next task uses the when condition when calling the ansible.builtin.import_tasks module, first of all for the Debian-based system:

```
- name: "Install the stack on Debian based systems"
  ansible.builtin.import_tasks: "Debian.yml"
  when: ansible_os_family == 'Debian'
```

In our case, this means that when the Ansible playbook is targeting a Debian-based host, it will load the tasks from roles/stack_install/tasks/Debian.yml, which are essentially the same as those we discussed at length in *Chapter 5, Deploying WordPress*, and execute them against the host.

The next task does the same function, but this time for Red Hat-based hosts, using the tasks listed in the roles/stack_install/tasks/RedHat.yml file:

```
- name: "Install the stack on RedHat based systems"
  ansible.builtin.import_tasks: "RedHat.yml"
  when: ansible_os_family == 'RedHat'
```

The roles/stack_install/tasks/RedHat.yml file contains three tasks, which are pretty much the same as the Debian.yml tasks.

We start the role by running an update of all the installed packages:

```
- name: "Update all of the installed packages"
  ansible.builtin.dnf:
    name: "*"
    state: "latest"
    update_cache: true
```

As you can see, this uses the `ansible.builtin.dnf` modules rather than the `ansible.builtin.apt` one.

Next up, we have the task that installs the NGINX mainline repo:

```
- name: "Add the NGINX mainline repo"
  ansible.builtin.yum_repository:
    name: "{{ nginx_repo.name }}"
    description: "{{ nginx_repo.description }}"
    baseurl: "{{ nginx_repo.baseurl }}"
    gpgcheck: "{{ nginx_repo.gpgcheck }}"
    enabled: "{{ nginx_repo.enabled }}"
```

Although this uses the `ansible.builtin.yum_repository` module, DNF will pick up the new repo once it is added. This is also the only task we need to run to add the repo, and adding a Yum repository is very different from adding a repository on a Debian-based system.

The final task for Red Hat-based systems is to install all the packages, including the NGINX one from the mainline repository we just enabled by, again, calling the `ansible.builtin.dnf` module:

```
- name: "Update cache and install the stack packages"
  ansible.builtin.dnf:
    state: "present"
    update_cache: true
    pkg: "{{ system_packages + extra_packages + stack_packages }}"
```

As you can see, with a little change to the logic in which the tasks are being called, it was relativity painless to update the role to target Debian and Red Hat distributions.

For the next role we need to change, the Stack Config role, we will take a slightly different approach to considering the different operating system distributions.

## The Stack Config role

Apart from a single task at the start and half a dozen at the end, the bulk of this role remains as-is.

There are some changes to the default variables file in the `roles/stack_config/default/main.yml` file; first off, the following variables are added:

```
selinux:
  http_permissive: true

firewall_comands:
  - "firewall-cmd --zone=public --add-port=80/tcp --permanent"
  - "firewall-cmd --zone=public --add-port=80/tcp"
```

As I am sure you can guess from their names, these deal with SELinux and the Firewall.

The next change is to move the `mysql_socket_path`, `php_fpm_path`, `php_ini_path`, and `php_service_name` variables to distribution-specific files at `roles/stack_config/vars/Debian.yml` and `roles/stack_config/vars/RedHat.yml`.

As we have already discussed, one of the key differences between the two distributions is the paths to both the core files and the configuration files for the services we installed during the Stack Install role.

In the `roles/stack_config/vars/Debian.yml` file, we have the following:

```
mysql_socket_path: "/var/run/mysqld/mysqld.sock"
php_fpm_path: "/etc/php/8.1/fpm/pool.d/www.conf"
php_ini_path: "/etc/php/8.1/fpm/php.ini"
php_service_name: "php8.1-fpm"
```

However, for the `roles/stack_config/vars/RedHat.yml` file, we need to define the following:

```
mysql_socket_path: "/var/lib/mysql/mysql.sock"
php_fpm_path: "/etc/php-fpm.d/www.conf"
php_ini_path: /etc/php.ini
php_service_name: "php-fpm"
```

As you can see, at first glance, they look a little similar, but the paths and file names are different.

These files are called by a task, which is the same as we used at the start of the Stack Install role:

```
- name: Include the operating system specific variables
  ansible.builtin.include_vars: "{{ ansible_os_family }}.yml"
```

From here, all the original tasks we covered in *Chapter 5, Deploying WordPress*, are called and executed, ending with the task that removes the test MySQL database.

From here, in the role, we have the tasks that consider the additional steps needed to configure our Red Hat-based host, starting with configuring SELinux; for our role, we need to enable the policy that allows web servers to run, for which, on a lot of Red Hat distributions, is blocked by default.

The task to do this looks like the following:

```
- name: "Set the selinux allowing httpd_t to be permissive is
required"
  community.general.selinux_permissive:
    name: httpd_t
    permissive: true
  when: selinux.http_permissive and ansible_os_family == 'RedHat'
```

As you can see, the when condition here ensures that the task is only executed when the `selinux.http_permissive` variable is set to `true`, and the `ansible_os_family` is equal to RedHat.

While our Debian-based system will meet the `selinux.http_permissive` condition, the task will be skipped on those hosts, as it doesn't meet the second condition.

Finally, we have the tasks for configuring the `firewalld` service, which is the default firewall on most modern Red Hat-based distributions.

> **Important note**
>
> Although we are using `firewall-cmd` in this section, there is an Ansible module that supports the firewalld service called `ansible.posix.firewalld`. As this is the only instance in the title that we will be targeting a Red Hat-based operating system with, we have, instead, used `ansible.builtin.command` to show how we can meet more complex conditions based on the output commands.

Like some of the roles we have included in other chapters, configuring the firewall is a task we only have to do once. The first thing we will do is check for the presence of a file at `~/firewall-configured` and register the results:

```
- name: "Check to see if the ~/firewall-configured file exists"
  ansible.builtin.stat:
    path: "~/firewall-configured"
  register: firewall_configured
```

Next, we need to check whether `firewalld` is running, but only if it's a RedHat distribution. To do this, we need to run the `firewall-cmd --state` shell command and the output result is registered in the `fireweall_status` variable:

```
- name: "Check if firewalld is running"
  ansible.builtin.command: firewall-cmd --state
```

```
  register: fireweall_status
  when: ansible_os_family == 'RedHat'
```

Now, as the remaining tasks could also be executed on a Debian-based host, we need to take that into account as we now have a variable containing the stdout of the command we ran called fireweall_status, which won't be present, resulting in an error that would stop playbook execution on a Debian-based host:

```
- name: "Set a fact so the playbook can continue if running on a
Debian based system"
  ansible.builtin.set_fact:
    fireweall_status:
      stdout: notrunning
  when: ansible_os_family == 'Debian'
```

As you can see from the preceding task, if ansible_os_family is Debian, we are setting the fireweall_status.stdout variable to notrunning.

Now we have all of the information we need to make a decision on whether we should run the commands to configure the firewall, the following conditions need to be met:

- The firewall-cmd --state command returns running

- The operating system is RedHat

- The ~/firewall-configured file does not exist

If all three of these conditions are met, which are defined in the following task, then the commands to configure the firewall to open and allow traffic on port 80 are executed:

```
- name: "Run the firewall-cmd commands if the firewall-cmd --state
command returns running"
  ansible.builtin.command: "{{ item }}"
  with_items: "{{ firewall_comands }}"
  when: fireweall_status.stdout == "running" and ansible_os_family ==
'RedHat' and not firewall_configured.stat.exists
```

The final task then creates the ~/firewall-configured file so that the commands are not executed again:

```
- name: "Create the ~/firewall-configured file"
  ansible.builtin.file:
    path: ~/firewall-configured
    state: touch
    mode: "0644"
  when: not firewall_configured.stat.exists
```

It does this on both distributions, as it doesn't matter if it is set on Debian-based systems, and we don't want to run the commands regardless; on Red Hat systems, it will mean that any subsequent executions of the playbook will not be able to meet the three conditions where the commands are executed to configure the firewall service.

## The WordPress role

As already mentioned, we do not have to make any changes to this role.

# Running the playbook

There are no changes to our `site.yml` file, meaning that we need to run the following command to run the playbook:

```
$ ansible-playbook -i hosts site.yml
```

There is way too much output to cover here, but I will include some of the highlights from the playbook execution, starting with the gathering of the facts:

```
TASK [Gathering Facts] **********************************
ok: [Debian]
ok: [RedHat]
```

Now that Ansible knows about our two hosts, it makes a start on running the tasks; here are the updated ones from the Stack Install role:

```
TASK [roles/stack_install : update apt-cache and upgrade packages]
*************
skipping: [RedHat]
changed: [Debian]
```

As you can see, this was the `apt` one, and the `dnf` one looks like this:

```
TASK [roles/stack_install : update all of the installed packages]
*************
skipping: [Debian]
changed: [RedHat]
```

Now, moving onto the Stack Config role, this is where tasks are being run on both distributions:

```
TASK [roles/stack_config : add the wordpress group] *******
changed: [RedHat]
changed: [Debian]
```

To update the firewall on just the Red Hat-based distribution, we do the following:

```
TASK [roles/stack_config : run the firewall-cmd commands if the
firewall-cmd --state command returns running] ***
skipping: [Debian] => (item=firewall-cmd --zone=public --add-port=80/
tcp --permanent)
skipping: [Debian] => (item=firewall-cmd --zone=public --add-port=80/
tcp)
skipping: [Debian]
changed: [RedHat] => (item=firewall-cmd --zone=public --add-port=80/
tcp --permanent)
changed: [RedHat] => (item=firewall-cmd --zone=public -
-add-port=80/tcp)
```

Finally, we complete the playbook run:

```
PLAY RECAP ************************************************
Debian                     : ok=44    changed=29    unreacha-
ble=0    failed=0    skipped=7    rescued=0    ignored=2
RedHat                     : ok=45    changed=34    unreacha-
ble=0    failed=0    skipped=6    rescued=0    ignored=2
```

All of which means that I should now have two WordPress installations:

Figure 6.4 – WordPress running on Ubuntu 22.04

Figure 6.5 – WordPress running on Rocky Linux 9.2

While the preceding screens aren't the most exciting of websites, as you can see, we have WordPress up and running on two different operating systems.

At this point, if you have been following along, don't forget to delete any resources you have deployed to run your playbooks against.

# Summary

In this chapter, we have adapted the WordPress installation playbook we wrote in *Chapter 5, Deploying WordPress*, to target multiple operating systems. We did this by using Ansible's built-in auditing module to determine which operating system the playbook is running against and running only the tasks that will work on the two target distributions.

While targeting multiple Linux distributions is one use for the approach we have taken with the conditions we have been using, I am sure that you will already have some ideas on how you could use some of the logic we used in your projects, such as bootstrapping different software based on the role on a virtual machine host, etc.

Additionally, this approach is beneficial when publishing your roles to Ansible Galaxy, as discussed in *Chapter 2, Exploring Ansible Galaxy*, by making the operating system agnostic.

You may have noticed so far that we have been targeting Linux virtual machines; in the next chapter, we will look at Ansible support for Windows-based operating systems.

# Further reading

- **Red Hat Enterprise Linux**: https://www.redhat.com/en/technologies/linux-platforms/enterprise-linux
- **Debian**: https://www.debian.org/
- **Ubuntu**: https://ubuntu.com/
- **Rocky Linux**: https://rockylinux.org/
- **The Register**, *Red Hat strikes a crushing blow against RHEL downstreams*: https://www.theregister.com/2023/06/23/red_hat_centos_move/
- **The Ansible Posix Firewalld Module**: https://docs.ansible.com/ansible/latest/collections/ansible/posix/firewalld_module.html

# 7

# Ansible Windows Modules

In this chapter, we will look at the ever-growing collection of built-in Ansible modules that support and interact with Windows-based servers; coming from an almost exclusively macOS and Linux background, it seemed odd to be using a tool not natively supported on Windows to manage Windows.

By the end of our time in this chapter, I am sure you will agree that looking at the options available, the Ansible developers have made managing Windows Server workloads with a playbook as seamless and familiar as possible.

In this chapter, we will learn how to do the following:

- Launch a Windows server instance in Microsoft Azure
- Enable features in Windows
- Create users
- Install third-party packages using Chocolatey

The chapter covers the following topics:

- Launching a Windows server in Azure
- Ansible preparation
- The Windows Playbook roles
- Running the Playbook

## Technical requirements

Rather than trying to run a full Windows Server 2022 locally in a **virtual machine** (**VM**), in this chapter, we will cover securely launching and configuring a Windows Server 2022 VM hosted in Microsoft Azure. If you are following along, you must have an active Microsoft Azure subscription and the Azure **command-line interface** (**CLI**) installed.

For details on how to install and configure the Azure CLI, please see the documentation at `https://learn.microsoft.com/en-us/cli/azure/install-azure-cli/`. If you are following along on a Windows host, I recommend installing the Azure CLI within your Windows Subsystem for Linux installation alongside where you installed Ansible.

## Launching a Windows server in Azure

We will not use Ansible to deploy the Azure resources as we will do in *Chapter 9, Moving to the Cloud*; instead, we will use the Azure CLI to launch our VM.

> **Note**
>
> As some of the commands in this chapter will be pretty long, I will break them up with a backslash. In Linux command lines, the backslash (\) followed by a newline is a line continuation character. It lets you split a single command over multiple lines for better readability.

Start by changing to the `Chapter07` folder within your checked-out copy of the repository that accompanies this title and run the following commands:

```
$ MYIP=$(curl https://api.ipify.org 2>/dev/null)
$ VMPASSWORD=$(openssl rand -base64 24)
$ echo $VMPASSWORD > VMPASSWORD
```

The first two commands set two variables on your command line; the first uses the **ipify** service (`https://www.ipify.org/`) to populate the $MYIP variable with the public IP address of your current network session.

The second generates a random password using the `openssl` command and assigns it to the variable called $VMPASSWORD.

The third command copies the content of $VMPASSWORD to a file called VMPASSWORD; this command must be executed in the same folder as the host inventory file, as it will be called in our host inventory file, which we will discuss later in the chapter.

> **Note**
>
> I will follow the Azure Cloud Adoption Framework recommendations around resource naming and launching the resources in the UK South region.

Now that we know our IP address and have a password, we can start using the Azure CLI to launch resources. The first thing we need to do is make sure that we are logged in by running the following:

```
$ az login
```

Once logged in, we can then create an **Azure Resource Group** by executing the following:

```
$ az group create \
    --name rg-ansible-windows-server-uks \
    --location uksouth
```

The Azure Resource Group is the logic container we will be deploying our Azure resources to, the first of which will be an **Azure Virtual Network**.

The following command will create the Azure Virtual Network with an address space of 10.0.0.0/24 and a single subnet using 10.0.0.0/27; this is where we will launch our Windows Server:

```
$ az network vnet create \
    --resource-group rg-ansible-windows-server-uks \
    --name vnet-ansible-windows-server-uks \
    --address-prefix 10.0.0.0/24 \
    --subnet-name sub-vms \
    --subnet-prefix 10.0.0.0/27
```

Now, we need to create a **Network Security Group** to assign to the network interface of our VM once it has been launched.

We need this as we will assign a public IP address to the VM, and we don't want to expose our three management ports directly to the internet; instead, we want to lock them down to just us:

```
$ az network nsg create \
    --resource-group rg-ansible-windows-server-uks \
    --name nsg-ansible-windows-server-uks
```

We now have an empty Network Security Group created. Let's add some rules, starting with the rule that opens port 80 to everyone to allow HTTP traffic:

```
$ az network nsg rule create \
    --resource-group rg-ansible-windows-server-uks \
    --nsg-name nsg-ansible-windows-server-uks \
    --name allowHTTP \
    --protocol tcp \
    --priority 100 \
    --destination-port-range 80 \
    --access allow
```

Next, we have the rule that opens port 3389, which **Remote Desktop** uses to allow you to create a session to the host; we only want this open to us, so the command here would be as follows:

```
$ az network nsg rule create \
    --resource-group rg-ansible-windows-server-uks \
```

```
    --nsg-name nsg-ansible-windows-server-uks \
    --name allowRDP \
    --protocol tcp \
    --priority 1000 \
    --destination-port-range 3389 \
    --source-address-prefix $MYIP/32 \
    --access allow
```

Note that we are passing in the $MYIP variable we registered when launching the resources. This will pass your IP address, and as you can see, we are then appending /32 to the end; this is the **Classless Inter-Domain Routing (CIDR)** notation for a single IP address.

Now that we have the rule for Remote Desktop in place, which is how we, as end users, will connect to the VM, we need to open the **Windows Remote Management (WinRM)** port, which is how Ansible will be connecting to the machine:

```
$ az network nsg rule create \
    --resource-group rg-ansible-windows-server-uks \
    --nsg-name nsg-ansible-windows-server-uks \
    --name allowWinRM \
    --protocol tcp \
    --priority 1050 \
    --destination-port-range 5985-5986 \
    --source-address-prefix $MYIP/32 \
    --access allow
```

The next of the commands we need to run is the one that launches the VM itself and configures it to use the core networking components we have just launched:

```
$ az vm create \
    --resource-group rg-ansible-windows-server-uks \
    --name vm-ansible-windows-server-uks \
    --computer-name ansibleWindows \
    --image Win2022Datacenter \
    --admin-username azureuser \
    --admin-password $VMPASSWORD \
    --vnet-name vnet-ansible-windows-server-uks \
    --subnet sub-vms \
    --nsg nsg-ansible-windows-server-uks \
    --public-ip-sku Standard \
    --public-ip-address-allocation static
```

As you can see, we are instructing the Azure CLI to launch a VM that uses the **Win2022Datacenter** VM image as its base; the VM is being deployed into the `rg-ansible-windows-server-uks` resource group and using all of the network resources we launched using the previous commands.

You might be thinking, great, let's get back to looking at Ansible. However, there is one more command we need to run before we can connect to the VM using Ansible – and the reason is that while we have a Windows 2022 server up and running, the WinRM protocol is not enabled by default.

The command to enable this functionality is as follows:

```
$ az vm extension set \
    --resource-group rg-ansible-windows-server-uks \
    --vm-name vm-ansible-windows-server-uks \
    --name CustomScriptExtension \
    --publisher Microsoft.Compute \
    --version 1.10 \
    --settings "{'fileUrls': ['https://raw.githubusercontent.
com/PacktPublishing/Learn-Ansible-Second-Edition/main/Scripts/
ConfigureRemotingForAnsible.ps1'],'commandToExecute': 'powershell
-ExecutionPolicy Unrestricted -File ConfigureRemotingForAnsible.ps1'}"
```

This enables a VM Extension on the Azure VM we have just deployed. There are many different types of Virtual Machine Extensions; the one we are using is **Custom Script Extension**. This extension downloads a script from a URL passed to it and then executes a command; in our case, we are downloading the script configuring WinRM from the GitHub repository accompanying this title.

You can see the script that will be downloaded at the following URL: `https://raw. githubusercontent.com/PacktPublishing/Learn-Ansible-Second-Edition/ main/Scripts/ConfigureRemotingForAnsible.ps1`

The command that runs once the script is downloaded is as follows:

```
$ powershell -ExecutionPolicy Unrestricted -File
ConfigureRemotingForAnsible.ps1
```

The Virtual Machine Extension executes the preceding command, so we do not have to run it directly.

The Resource Visualizer in the Azure portal for the resource group should show you something that looks like the following overview:

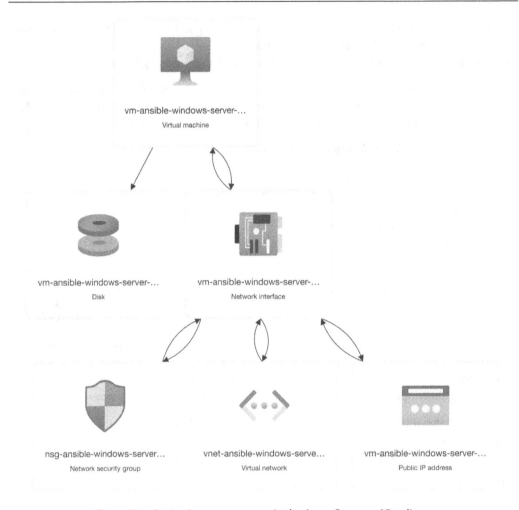

vm-ansible-windows-server-...
Virtual machine

vm-ansible-windows-server-...
Disk

vm-ansible-windows-server-...
Network interface

nsg-ansible-windows-server...
Network security group

vnet-ansible-windows-serve...
Virtual network

vm-ansible-windows-server-...
Public IP address

Figure 7.1 – Reviewing our resources in the Azure Resource Visualizer

Once completed, our Windows Server VM is ready to have our Ansible run against it.

## Ansible preparation

As mentioned in the previous section, Ansible will use WinRM to interact with our Windows host.

> **Information**
>
> WinRM provides access to a **Simple Object Access Protocol (SOAP)**-like protocol called **WS-Management**. Unlike **Secure Shell (SSH)**, which provides the user with an interactive shell to manage the host, WinRM accepts executed scripts, and the results are passed back to you.

Ansible requires us to install a few Python modules to enable it to use the protocol; these modules need to be installed as they are not typically installed alongside Ansible.

To install the module, if you are running on Ubuntu, run the following command:

```
$ sudo -H pip install pywinrm
```

On macOS, run the following:

```
$ pip install pywinrm
```

Once installed, we need to update our environment file to instruct Ansible to use the WinRM protocol rather than SSH.

Our updated `hosts` file looks like the following file, which is a copy of the `hosts.example` file from the `Chapter07` folder in the accompanying repository. If you are following along with the exercise, you will need to update yours to update the IP address to match that of your Azure Virtual Machine once it has been launched:

```
WindowsServer ansible_host=123.123.123.123.nip.io

[ansible_hosts]
WindowsServer

[ansible_hosts:vars]
ansible_connection=winrm
ansible_user="azureuser"
ansible_password="{{ lookup('ansible.builtin.file', 'VMPASSWORD') }}"
ansible_winrm_server_cert_validation=ignore
```

The start of the file mirrors what we have been used to so far in that it contains a name for a host and the resolvable hostname of the VM, again using the **Nip.io** service (`https://nip.io/`).

Next, we take the named host and put it in the `ansible_hosts` group before defining a bunch of settings for the group.

The first of these settings instructs Ansible to use `winrm` by setting it as the value for the `ansible_connection` key.

Next, we set the `ansible_user` key; the value is `azureuser`, which we defined when we launched the Azure Virtual Machine; and also the `ansible_password` key.

If you recall, at the start of the chapter, we ran the following command:

```
$ echo $VMPASSWORD > VMPASSWORD
```

This took the random password we generated, that is, $VMPASSWORD, and placed it inside a file named VMPASSWORD; this means that when we define the ansible_password key, we can use a lookup value, using {{ lookup('ansible.builtin.file', 'VMPASSWORD') }}, to read the contents of the VMPASSWORD file and use that rather than us having to hardcode the password into our environment file.

Finally, we tell Ansible to ignore any certificate errors by setting the ansible_winrm_server_cert_validation key to false; we need to do this because WinRM has been configured to use a self-signed certificate, which will cause a certificate error as our local machine does not know to trust the certificate.

Now that we have Windows up and running and Ansible configured, we can start interacting with it.

## The ping module

Not all Ansible modules work with Windows hosts, and ansible.builtin.ping is one of them.

If you were to run the following command:

```
$ ansible WindowsServer -i hosts -m ansible.builtin.ping
```

You would then get quite a verbose error with the following warning:

```
[WARNING]: No python interpreters found for host WindowsServer (tried
['python3.11', 'python3.10',
'python3.9', 'python3.8', 'python3.7', 'python3.6', 'python3.5', '/
usr/bin/python3',
'/usr/libexec/platform-python', 'python2.7', '/usr/bin/python',
'python'])
```

Luckily, there is a module provided for Windows called ansible.windows.win_ping, so let's update our command to run that instead:

```
$ ansible WindowsServer -i hosts -m ansible.windows.win_ping
```

This returns the result you would expect to receive if you sent a ping:

```
WindowsServer | SUCCESS => {
    "changed": false,
    "ping": "pong"
}
```

The next module we will look at doesn't require any changes from how we ran it against a Linux host.

## The setup module

As before, we need to run the module and target our host, so the command is as follows:

```
$ ansible WindowsServer -i hosts -m ansible.builtin.setup
```

This will return information on the host as it did when executing the same module against our Linux host, a snippet of which can be seen in the following screenshot:

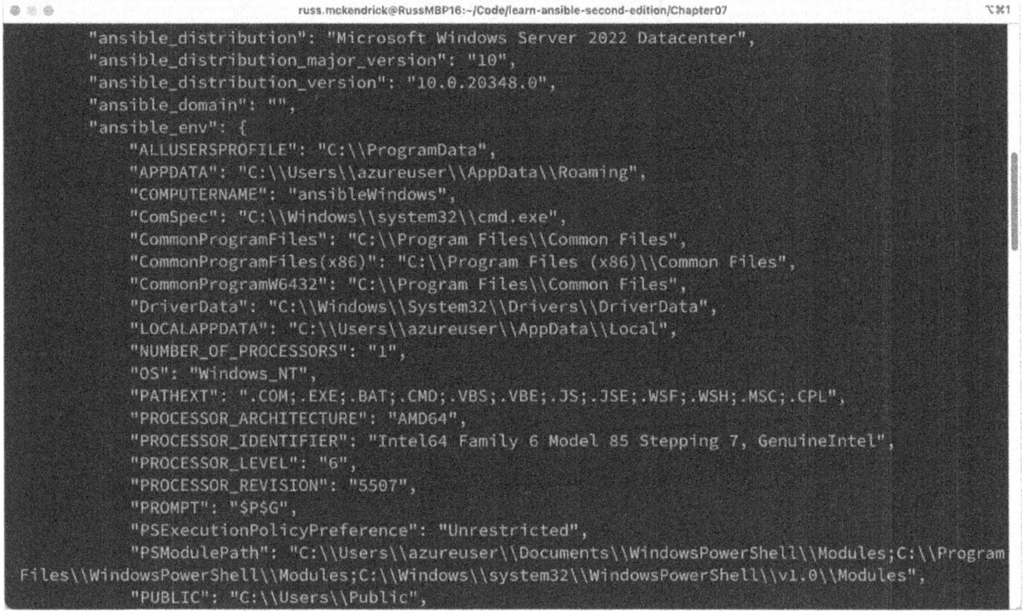

Figure 7.2 – Some of the output from the setup module

This is one of the only modules that will work on Windows and Linux hosts.

Now that we have confirmed that our host is accessible, let's look at the changes we need to make to the playbooks.

# The Windows Playbook roles

The entire playbook can be found in the Chapter 07 folder in the repository that accompanies the title, so I will not cover how to create roles in this chapter as we have covered it at length in the previous chapters.

## Enabling Windows features

Two roles cover how to enable features; the first role, called iis, enables the **Internet Information Services (IIS)** on our Windows Server.

> **Information**
>
> IIS is the default web server that ships with Windows Server, and it supports the following protocols: HTTP, HTTPS, and HTTP/2, as well as FTP, FTPS, SMTP, and NNTP. It was first released in 1995 as part of Windows NT.

There are some default variables in roles/iis/defaults/main.yml; these define where things need to be copied to on the server and also include the contents on an HTML file we will copy to the host:

```
document_root: 'C:\inetpub\wwwroot\'
html_file: "ansible.html"
html_heading: "Success !!!"
html_body: |
  This HTML page has been deployed using Ansible to a <b>{{ ansible_
distribution }}</b> host.<br><br>
  The weboot is <b>{{ document_root }}</b> this file is called <b>{{
html_file }}</b>.<br>
```

There are then two tasks in roles/iis/tasks/main.yml. The first task is *where the magic happens*:

```
- name: "Enable IIS"
  ansible.windows.win_feature:
    name:
      - "Web-Server"
      - "Web-Common-Http"
    state: "present"
```

I say *where the magic happens* because I don't get to touch Windows hosts very often as a Linux system administrator by trade.

Still, as you can see from the preceding task, Ansible is giving us a Linux-like experience, meaning that I don't have to roll up my sleeves and get under the hood of Windows too much.

The task uses the ansible.windows.win_feature module to enable the Web-Server and Web-Common-Http features; as we are sticking with the default out-of-the-box settings, there isn't any more configuration we need to do other than to copy an HTML file across to the document root:

```
- name: "Create an html file from a template"
  ansible.windows.win_template:
    src: "index.html.j2"
    dest: "{{ document_root }}{{ html_file }}"
```

As you can see, we are using a Jinja2 template file, an abridged version of which looks like the following:

```
<!--{{ ansible_managed }}-->
<title>{{ html_heading }}</title>
<article>
    <h1>{{ html_heading }}</h1>
    <div>
        <p>{{ html_body }}</p>
    </div>
</article>
```

But rather than `ansible.builtin.template`, we are using `ansible.windows.win_template`, which is the Windows module version, as I am sure you will have already guessed.

Suppose we were to use the `ansible.builtin.template` version; we would get the same error as when we ran the `ansible.builtin.ping` module, and complaints about Python not being installed.

The next role expands on the `iis` file and enables `.Net`; the role is called `dotnet`.

Again, there are some default variables in `roles/dotnet/defaults/main.yml`:

```
aspx_document_root: 'C:\inetpub\wwwroot\ansible\'
aspx_file: "default.aspx"
aspx_heading: "Success !!!"
aspx_body: |
  This HTML page has been deployed using Ansible to a <b>{{ ansible_
distribution }}</b> host.<br><br>
  The weboot is <b>{{ aspx_document_root }}</b> this file is called
<b>{{ aspx_file }}</b>.<br><br>
  The output below is from ASP.NET<br><br>
  Hello from <%= Environment.MachineName %> at <%= DateTime.UtcNow
%><br><br>
```

As you can see, this time, the body contains some inline code.

However, you may have yet to spot a subtle difference in how we define the paths in the variables. For both tasks for our Windows workload, the path variables have been defined as follows:

```
document_root: 'C:\inetpub\wwwroot\'
aspx_document_root: 'C:\inetpub\wwwroot\ansible\'
```

But, if we look at how we defined the path in *Chapter 5, Deploying WordPress*, there is quite a crucial difference:

```
wordpress_system:
    home: "/var/www/wordpress"
```

The difference is not that we have used `wordpress_system.home` as the variable; it is more subtle than that.

If you have noticed that the Windows workload paths are using single quotes and the Linux one is using double quotes, give yourself a pat on the back.

In Ansible, single quotes ( ' ) that enclose strings are treated as literals, ensuring special characters aren't interpreted or expanded, making them ideal for Windows paths.

Double quotes ( " ) allow for string interpolation, meaning embedded Jinja2 template expressions or special characters will be expanded. They also support escape sequences, such as \n for new lines, because many escape sequences such as \, which is in our path, could cause problems.

If we needed to use double quotes because we needed to pass in something that needed to be expanded, then you could have a double slash ( \\ ) like this:

```
document_root: "C:\\inetpub\\wwwroot\\"
aspx_document_root: "C:\\inetpub\\wwwroot\\ansible\\"
```

However, it can confuse reading the paths, so I used single quotes in our examples – back to the role now.

The first of four tasks in `roles/dotnet/tasks/main.yml` enables `.Net`:

```
- name: "Enable .NET"
  ansible.windows.win_feature:
    name:
      - "Net-Framework-Features"
      - "Web-Asp-Net45"
      - "Web-Net-Ext45"
    state: "present"
  notify: "Restart IIS"
```

We are also triggering a restart of IIS via a handler if any changes are detected; this uses `ansible.windows.win_service`:

```
- name: "Restart IIS"
  ansible.windows.win_service:
    name: "w3svc"
    state: "restarted"
```

The next task creates a folder if one doesn't exist:

```
- name: "Create the folder for our asp.net app"
  ansible.windows.win_file:
    path: "{{ aspx_document_root }}"
    state: "directory"
```

Again, a Windows version of an existing module we have used is called, this time `ansible.windows.win_file`. Next, we copy the file to the folder we just created:

```
- name: "Create an aspx file from a template"
  ansible.windows.win_template:
    src: "default.aspx.j2"
    dest: "{{ aspx_document_root }}{{ aspx_file }}"
```

The final task in the role configures IIS to consider we are now running an application:

```
- name: "Ensure the default web application exists"
  community.windows.win_iis_webapplication:
    name: "Default"
    state: "present"
    physical_path: "{{ aspx_document_root }}"
    application_pool: "DefaultAppPool"
    site: "Default Web Site"
```

There are a few more roles to cover before we run the playbook; let's look at the next one.

## Creating a user

This role creates a user for us to connect to our instance with. The defaults that can be found in `roles/user/defaults/main.yml` are as follows:

```
ansible:
  username: "ansible"
  password: "{{ lookup('password', 'group_vars/generated_password
chars=ascii_letters,digits length=30') }}"
  groups:
    - "Users"
    - "Administrators"
```

As you can see, here, we are defining a user called `ansible` that has a 30-character random password, which Ansible will create using a lookup plugin if one doesn't exist. The `ansible` user will be a member of the `Users` and `Administrators` groups.

There is a single task in `roles/user/tasks/main.yml` using the `ansible.windows.win_user` module, which looks like the following:

```
- name: "Ensure that the ansible created users are present"
  ansible.windows.win_user:
    name: "{{ ansible.username }}"
    fullname: "{{ ansible.username | capitalize }}"
    password: "{{ ansible.password }}"
```

```
    state: "present"
    groups: "{{ ansible.groups }}"
```

Like all Windows modules, the syntax is similar to the Linux equivalent, so you should know what each key means. As you can see from the previous task, we are using a Jinja2 transformation to capitalize the first letter of the `ansible.username` variable.

## Installing applications using Chocolatey

The next role, called `choco`, uses **Chocolatey** to install a few bits of software on the machine.

> **Information**
>
> Chocolatey is Windows' answer to macOS's Homebrew – a package manager streamlining software installations. Like we used Homebrew earlier, Chocolatey wraps typical Windows installations into neat PowerShell commands, making it a perfect match for orchestration tools such as Ansible.

In `roles/choco/defaults/main.yml`, we have a single variable that contains a list of the packages we want to install:

```
apps:
    - "notepadplusplus.install"
    - "putty.install"
    - "googlechrome"
```

As you may have already guessed, this is the task that installs the applications:

```
- name: "Install software using chocolatey"
  chocolatey.chocolatey.win_chocolatey:
    name: "{{ apps }}"
    state: "present"
```

Again, the module takes a similar input to the previous package manager modules, `ansible.builtin.apt` and `ansible.builtin.dnf`, that we used. This means that the logic Ansible uses across the modules that do similar tasks is consistent across multiple operating systems and not just different Linux distributions.

## Information role

The final role is called info; its only purpose is to output once the playbook has finished running. The role has a single task defined in `roles/info/tasks/main.yml`:

```
- name: "Print out information on the host"
  ansible.builtin.debug:
    msg: "You can connect to '{{ ansible_host }}' using the username
of '{{ ansible.username }}' with a password of '{{ ansible.password
}}'."
```

As you can see, this will provide us with the hostname to create a Remote Desktop session, along with confirming the username and password we should use.

That concludes our look at the roles that will be called when we run the playbook, which we are now ready to do.

## Running the Playbook

The `site.yml` is missing some of the settings at the top because we are targeting a Windows host:

```
---
- name: "Install IIS, .NET, create user, install chocolatey and
display info"
  hosts: "ansible_hosts"
  gather_facts: true
  vars_files:
    - "group_vars/common.yml"

  roles:
    - "iis"
    - "dotnet"
    - "user"
    - "choco"
    - "info"
```

As you can see, there is no need for the become or become_method keys to be set, as we do not need to change users once connected to the host.

Outside of that, the rest of the file is as expected, as is the way we run the playbook:

```
$ ansible-playbook -i hosts site.yml
```

It will take a little while to run as a lot is going on in the background, as you will see from the output when the playbook runs for the first time:

```
                    russ.mckendrick@RussMBP16:~/Code/learn-ansible-second-edition/Chapter07                    ⌥⌘1
changed: [WindowsServer]

TASK [roles/dotnet : create an aspx file from a template] ***************************************
changed: [WindowsServer]

TASK [roles/dotnet : ensure the default web application exists] ***********************************
changed: [WindowsServer]

TASK [roles/user : ensure that the ansible created users are present] *****************************
changed: [WindowsServer]

TASK [roles/choco : install software using chocolatey] ****************************************
[WARNING]: Chocolatey was missing from this system, so it was installed during this task run.
changed: [WindowsServer]

TASK [roles/info : print out informaton on the host] *****************************************
ok: [WindowsServer] => {
    "msg": "You can connect to '20.50.120.120.nip.io' using the username of 'ansible' with a passwor
d of 'AejmAiTx2otFqQzKI9G3OGUubhtQjx'."
}

PLAY RECAP **************************************************************************************
WindowsServer                 : ok=10    changed=5    unreachable=0    failed=0    skipped=0    rescued=
0    ignored=0

~/Code/learn-ansible-second-edition/Chapter07  ⎇ main
```

Figure 7.3 – Reviewing the playbook output

As you can see from the preceding output, the host I was given was `20.50.120.120.nip.io` (this host has long since been terminated, but if you are following along, you can replace the preceding host with your own).

To view the static HTML and `.Net` pages we uploaded, you can visit `http://20.50.120.120.nip.io/ansible.html` or `http://20.50.120.120.nip.io/ansible/default.aspx`, making sure to update the host to reflect your own.

You can also open a remote desktop session to the host using the credentials given in the output; the following screenshot shows a session using the user we created and opening the side using **Google Chrome** with notes in **Notepad++**, both of which are applications we installed with the Playbook:

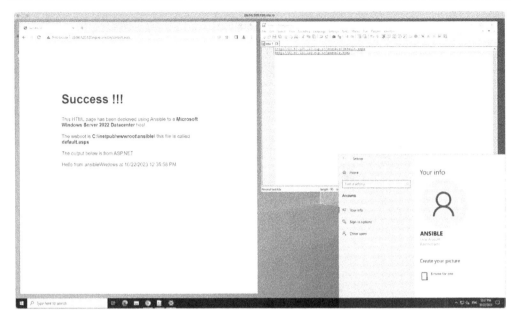

Figure 7.4 – A remote desktop session

Once you have finished with the host, you can run the following Azure CLI command to terminate all the resources we created:

```
$ az group delete \
    --name rg-ansible-windows-server-uks
```

Double-check that everything has been removed as expected to ensure you do not get any unexpected bills.

## Summary

As mentioned at the start of the chapter, using what we would consider a traditional Linux tool such as Ansible on Windows always feels a little strange. However, I am sure you will agree that the experience is as Linux-like as possible.

When I first started experimenting with the Windows modules, I was surprised that I managed to launch a Windows Server in Azure and deploy a simple web application without having to remote desktop into the target instance.

With each new release, Ansible gets more and more support for Windows-based hosts, making it easy to manage mixed workloads from your playbooks.

In the next chapter, we will examine the networking modules available in Ansible.

## Further reading

You can find more information on the excellent Chocolatey at `https://chocolatey.org/`.

# Part 3: Network and Cloud Automation

Ansible's power extends beyond just managing servers; it can also automate network devices and cloud infrastructure. Here, we will explore Ansible's network modules and discuss how to interact with network devices programmatically. We will then shift our attention to the cloud, where you will discover how to provision and manage resources on popular cloud platforms such as Microsoft Azure and Amazon Web Services. By the end of this part, you will have the skills needed to automate complex cloud deployments using Ansible.

This part has the following chapters:

- *Chapter 8, Ansible Network Modules*
- *Chapter 9, Moving to the Cloud*
- *Chapter 10, Building Out a Cloud Network*
- *Chapter 11, Highly Available Cloud Deployments*
- *Chapter 12, Building Out a VMware Deployment*

# 8

# Ansible Network Modules

Welcome to *Chapter 8*; this chapter will look at an often-overlooked use case for Ansible and delve into Ansible's expansive community of networking modules.

In this chapter, instead of a deep dive into every collection—which could be an entire book—we'll provide an overview, highlighting these modules' capabilities and flexibilities.

The chapter covers the following topic:

- Manufacturer and device support

## Manufacturer and device support

So far, we have been looking at modules that interact with servers. In our case, they have mostly been running locally. In the upcoming chapters, we will be communicating more with remotely cloud-hosted servers. But, before interacting with remote servers, we should cover the core network modules.

These modules have all been designed to interact with and manage the configuration of various network devices, from your traditional top-of-rack switches and fully virtualized network infrastructure to firewalls and load balancers. Ansible supports many devices, from open source virtual appliances to hardware solutions, some of which could have a starting price of over USD 500,000, depending on your configuration.

So, what do all these collections and modules have in common?

Well, they all interact with what are traditionally complex to configure devices, which, in most deployments and environments, are both the core and critical elements; after all, everything connected to them needs some level of network connectivity.

These modules give you a standard interface, i.e., Ansible, for many of these devices and remove the need for engineers to access these devices directly. They can, instead, run Ansible playbooks, which run roles created by experienced network engineers to configure them in a controlled and consistent way by just changing a few variables.

The only downside of using Ansible to manage this critical core infrastructure is that the host running Ansible requires a line of sight to the management interface or API running on the devices, which may sometimes raise some security concerns. Hence, Ansible's instructions on how to manage your network devices need some serious thought.

## The collections

The following list of collections is in the order of the namespaces and then the collection name; this is listed at the end of each item as **[namespace.collection-name]**.

### Apstra Extensible Operating System (EOS) [arista.eos]

There are over 30 modules that allow you to manage your devices running EOS. These modules let you operate **access control lists** (**ACLs**) interfaces, configure **border gateway protocol** (**BGP**) settings, run arbitrary commands on devices, manage hostname, interface configurations, logging, and more. A module also allows you to gather facts from each device.

Additionally, there are plugins for command line and HTTP API interactions.

### Check Point [check_point.mgmt]

The Ansible collection for Check Point Management comprises many modules; at the time of writing, there are over 250 modules.

Each manages different aspects of your Check Point device, such as access layers, rules, administrators, or network feeds on your Check Point firewall using the Web Services API. They provide functionalities ranging from fetching facts and adding or managing objects to workflow features such as approving and publishing sessions on your Check Point firewall.

### Cisco

Given the number of Cisco device types and classes, there are several collections in the Cisco namespace.

### Cisco Application Centric Infrastructure (ACI) [cisco.aci]

The 150+ ACI modules are used to manage all aspects of Cisco's ACI, which is to be expected of Cisco's next-generation API-driven networking stack.

There are modules for managing various aspects of the Cisco ACI, such as AAAA records (these are address records that store IPv6 addresses), roles, users, certificates, Access SPAN configurations, **bridge domains** (**BDs**) and subnets, BGP route reflectors, and many more. There are modules for managing Cloud Application Profiles and Cloud AWS Provider configurations.

### Cisco Adaptive Security Appliance (ASA) [cisco.asa]

With the five ASA modules, you can manage access lists, run commands, and manage the configuration of physical and virtual Cisco ASA-powered devices.

### Cisco DNA Center (DNAC) [cisco.dnac]

The Ansible collection for the Cisco DNAC comprises nearly 400 modules to manage different aspects of your Cisco DNAC deployment. The modules cover a range of functionalities, from fetching the configuration details of access points, managing application policies, assigning devices to sites, importing authentication certificates, and running compliance checks to managing configuration templates.

### Cisco IOS and IOS XR [cisco.ios and cisco.iosxr]

These two collections contain modules that allow you to manage your Cisco IOS and IOS XR-powered devices. You can gather facts on your devices and configure users, interfaces, logging, banners, and more with them.

### Identity Services Engine (ISE) [cisco.ise]

This collection manages your ISE; it comprises a variety of modules for managing settings and configurations, such as handling ACI bindings and settings, managing Active Directory settings, handling allowed protocols, administering ANC endpoints and policies, managing backup configurations and schedules, handling certificates, and more.

### Cisco Meraki [cisco.meraki]

Here, we have just short of 500 modules that manage the different elements of your Meraki deployment, such as administered identities, device details, camera settings, cellular gateway configurations, and sensor relationships. Each module is designed to fetch information or modify settings, which helps you manage your Cisco Meraki devices by using automation.

### Cisco Network Services Orchestrator (NSO) [cisco.nso]

A handful of modules allow you to interact with your Cisco NSO-managed devices. You can execute NSO actions, query data from your installation, and verify your configuration alongside service synchronization and configuration.

### Cisco Network Operating System Software (NX-OS) [cisco.nxos]

As you can imagine, there are a lot of modules for managing devices running Cisco NXOS; there are over 80 that cover a range of functions such as managing AAA server configurations, ACLs, BGP configurations, executing arbitrary commands, managing interfaces, and handling various other configurations and settings on Cisco NX-OS devices.

### Cisco Unified Computing System (UCS) [cisco.ucs]

While not strictly a networking device, the modules to manage Cisco's unified computing, storage, and network system include one that allows you to manage DNS servers, IP address pools, LAN connectivity policies, MAC address pools, QoS settings, VLANs, and vNICs. The rest of the modules allow you to programmatically manage computing and storage across your blades and chassis.

### *F5 BIG-IP Imperative [F5Networks.F5_Modules]*

There are 160 modules, all prefixed with BIG-IP, that allow you to manage all aspects of your F5 BIG-IP Application Delivery Controller.

### *Fortinet*

There are just two collections in the Fortinet namespace, but, as you can see from the number of modules in each, they are very feature-rich.

### Fortinet FortiManager [fortinet.fortimanager]

There are over 1,100 modules (yes, you read that correctly), including configuring antivirus profiles and options, managing AP local configuration profiles and command lists, configuring custom application signatures and firewall application groups, managing internet service applications, and more.

### Fortinet FortiOS v6 (fortinet.fortios)

While this has fewer modules than the FortiManager collection, there are still over 650 modules for configuring antivirus settings, application control lists, authentication schemes, and certificate settings.

### *Free Range Routing (FRR) [Frr.Frr]*

There are just two modules here: one that allows you to configure BGP, and the other lets you gather facts about devices running FRR.

### *Juniper Networks Junos [junipernetworks.junos]*

A total of 40 modules enable you to interact with Juniper devices running Junos from within your playbooks. These range from the standard command, configuration, and fact-gathering modules to those that allow you to install packages and copy files to your devices.

### *Open vSwitch [Openvswitch.Openvswitch]*

The four modules in the namespace allow you to manage bonds, bridges, ports, and databases on your OVS virtual switches.

## VyOS [vyos.vyos]

The VyOS collection includes modules for managing various configurations and resources on VyOS devices. Some of these modules include managing multiline banners, configuring BGP global and address family settings, running commands, managing firewall settings, interface configurations, logging, NTP, OSPF, SNMP, static routes, system commands, user management, and VLAN configurations, among others.

## The Community Network Collection [Community.Network]

This collection is a catch-all for all other network modules without dedicated namespace or development teams; the module prefix is now in the square brackets.

### A10 Networks [a10]

The A10 modules support A10 Networks AX, SoftAX, Thunder, and vThunder devices. These are all application delivery platforms that provide load balancing.

### Cisco AireOS [aireos]

The two AireOS modules allow you to interact with the Cisco Wireless LAN Controllers running AireOS. One of the modules will enable you to run commands directly on the devices, and the other is for managing the configuration.

### APCON [apcon]

A single module that allows you to run commands on your APCON device.

### Aruba Mobility Controller [aruba]

There are just two Aruba modules. These allow you to manage the configuration and execute commands on the Aruba Mobility Controllers from Hewlett Packard.

### Avi Networks [avi]

There are a total of 65 Avi modules that allow you to interact with all aspects of the Avi application services platform, including the load-balancing and **web application firewall (WAF)** features.

### Big Cloud Fabric and Big Switch Network [bcf + bigmon]

There are three Big Switch Network modules. **Big Cloud Fabric (BCF)** allows you to create and delete BCF switches. The other two modules enable you to create **Big Monitoring Fabric (Big Mon)** service chains and policies.

### Huawei Cloud Engine [ce]

Over 75 Cloud Engine modules allow you to manage all aspects of these robust switches from Huawei, including BGP, access control lists, MTU, static routes, VXLANs, and SNMP configuration.

### Lenovo CNOS [cnos]

There are nearly 30 modules that allow you to manage devices running the CNOS operating system from Lenovo; they enable you to configure everything from BGP and port aggregation to VLAG, VLANs, and factory reset devices, should you need to.

### Arista Cloud Vision [cv]

A single module lets you provision an Arista Cloud Vision server port using a configlet.

### illumos [dladm + flowadm + ipadm]

illumos is a fork of the Open Solaris operating system. Its powerful networking features make it the perfect candidate for deploying as a self-built router or firewall. These modules allow you to manage the interfaces, NetFlow, and tunnels. Additionally, as illumos is a fork of Open Solaris, your playbook should work on Open Solaris-based operating systems.

### Ubiquiti EdgeOS [edgeos + edgeswitch]

The modules for EdgeOS enable you to manage configurations, execute ad hoc commands, and collect facts on EdgeOS-running devices, such as the Ubiquiti Edge Router.

There are also a few modules for Edge Switches.

### Lenovo Enterprise Networking Operating System [enos]

There are three modules for the Lenovo ENOS. Like other devices, these allow you to gather facts, execute commands, and manage the configuration.

### Ericsson [eccli]

This single module allows you to run commands on devices running the Ericsson command-line interface.

### ExtremeXOS [exos + nos + slxos]

These half-a dozen modules allow you to interact with the ExtremeXOS, Extreme Networks SLX-OS, and Extreme Networks NOS software on Extreme Networks switches.

### Cisco Firepower Threat Defense [ftd]

A few modules allow you to configure and upload/download files to a Cisco Firepower Threat Defense device.

### Itential Automation Platform [iap]

A few modules allow you to interact with workflows hosted on the Itential Automation Platform, as well as low-code automation and orchestration for hybrid cloud networks.

### Ruckus ICX 7000 [icx]

These modules allow you to configure your Ruckus ICX 7000 series campus switches.

### Ingate Session Border Controllers [ig]

While these are mainly used for **SIP**, or to give its full name, **session initiation protocol** services, there are a few modules to help configure the network elements.

### NVIDIA Network Command Line Utility [nclu]

A single module that allows you to manage network interfaces using the NVIDIA Network Command Line Utility on compatible devices.

### Nokia NetAct [netact]

A single module that allows you to upload and apply your Nokia NetAct-powered core and radio networks.

### Citrix Netscaler [netscaler]

These modules are designed to manage and configure various aspects of Netscaler devices. They cover functionalities such as content switching, **Global Server Load-Balancing** (**GSLB**), load-balancing, issuing Nitro API requests, and saving configurations, as well as managing server configurations, services, service groups, and SSL certificate keys.

### Nokia Nuage Networks Virtualized Services Platform (VSP) [nuage]

There is a single module that allows you to manage enterprises on your Nokia Nuage Networks VSP.

### OpenSwitch [opx]

A single module that performs the specified action on the YANG object, utilizing the CPS API on the networking device operating OpenSwitch.

### Ordnance Virtual Routers [ordnance]

There are two modules: one to manage configuration and the other to collect facts on Ordnance Virtual Routers.

### *Pluribus Networks Netvisor OS [pn]*

These 40 modules allow you to manage your **Pluribus Networks (PN)** Netvisor OS-powered devices, from creating clusters and routers to running commands on your white-box switches.

### Nokia Networks Service Router Operating System [sros]

There are three modules that let you run commands against, configure, and roll back changes to your Nokia Networks SROS devices.

### Radware [vidrect]

A small number of modules that allow you to manage your Radware devices via a vDirect server.

### *Ansible Net Common [ansible.netcommon]*

The final collection is a set of modules that could be considered tools to help support all the devices we have covered in this chapter. There are modules that can ping targets and run generic commands using custom prompts and answers.

## Summary

I suspect most of you would not have heard of a lot of the devices we have listed in this chapter, and for the ones you have heard of—such as the Cisco ones—you will probably not have had direct access to them, leaving any configuration to your network administrators.

When we speak about triggering Ansible using CI/CD in *Chapter 15*, *Using Ansible with GitHub Actions and Azure DevOps*, and *Chapter 16*, *Introducing Ansible AWX and Red Hat Ansible Automation Platform*, we will learn about some deployment options that could help alleviate the concerns we mentioned at the start of the chapter, e.g., those about a host running your Ansible playbooks needing a line of sight of the potentially critical core infrastructure.

Before we get to those chapters, we will look at moving our workloads to the cloud, a journey that starts in the next chapter.

## Further reading

- **The Ansible Collection index**: https://docs.ansible.com/ansible/latest/collections/index.html

# 9

# Moving to the Cloud

This chapter will move from using our local virtual machine to using Ansible to launch instances with a public cloud provider.

For this chapter, we will be using Microsoft Azure, and we are targeting this provider as it allows us to launch virtual machines and interact with them without having too much configuration overhead.

We will then look at adapting our WordPress playbook to interact with the newly launched Microsoft Azure instance.

In this chapter, we will cover the following topics:

- An introduction to Microsoft Azure
- Launching instances in Microsoft Azure
- Bootstrapping WordPress

## Technical requirements

In this chapter, we will launch instances in a public cloud, so if you are following along, you will need a Microsoft Azure account. As with other chapters, complete versions of the playbooks can be found in the repository in the chapters folder at `https://github.com/PacktPublishing/Learn-Ansible-Second-Edition/tree/main/Chapter09`.

## An introduction to Microsoft Azure

In 2008, Microsoft took its first significant step into cloud computing by introducing Windows Azure, a cloud-based data center service. This launch marked a pivotal moment in what many people saw as a traditional software company's history, signaling a strategic shift toward cloud computing.

Developed under the internal project *Project Red Dog*, Windows Azure represented Microsoft's answer to the growing demand for scalable, accessible, and flexible computing resources.

Windows Azure was initially rolled out with five core components, each designed to offer distinct capabilities within the cloud computing spectrum:

- **Microsoft SQL Data Services**: This component offered a cloud version of Microsoft's SQL database, simplifying the complexities associated with hosting and managing databases in a cloud environment.

- **Microsoft .NET Services**: As a **platform as a service** (**PaaS**) offering, it enabled developers to deploy their .NET-based applications within a Microsoft-managed runtime, streamlining the development process.

- **Microsoft SharePoint** and **Microsoft Dynamics**: These **software as a service** (**SaaS**) offerings provided cloud-based versions of the company's renowned intranet and **customer relationship management** (**CRM**) products, enhancing collaboration and customer engagement.

- **Windows Azure (IaaS)**: An **infrastructure-as-a-service** (**IaaS**) solution, this allowed users to create and control virtual machines, storage, and networking services, addressing diverse compute workloads.

The preceding four definitions are from an older book I wrote, *Infrastructure as Code for Beginners*.

Central to Windows Azure's architecture was the Red Dog operating system, a specially modified version of Windows NT. This system was engineered to include a cloud layer, ensuring the smooth delivery of data center services.

By 2014, reflecting its expanded range of services and a growing emphasis on Linux-based workloads, Microsoft rebranded the service as Microsoft Azure. This change underscored the platform's evolution beyond Windows-centric solutions.

Fast forward to 2020, and it was evident that Microsoft Azure had embraced a more inclusive approach, with over half of its virtual machine cores and a significant number of Azure Marketplace images being Linux-based.

This shift demonstrated Microsoft's broader adoption of Linux and open-source technologies, which remain integral to their current cloud service offerings at the time of writing.

## Launching instances in Microsoft Azure

If you followed along in *Chapter 7, Ansible Windows Modules*, you will have already launched a virtual machine in Microsoft Azure using the Azure CLI.

> **Reminder**
>
> For instructions on how to install and configure the Azure CLI, please see the documentation at `https://learn.microsoft.com/en-us/cli/azure/install-azure-cli/`. Remember, if you are following along on a Windows host, then make sure to install the Azure CLI within your Windows Subsystem for Linux installation alongside where you installed Ansible.

When talking through launching the Windows virtual machine, we did the following:

- We created a resource group to collect all the resources for our virtual machine workload.

- We then created a virtual network and subnet to attach to the machine's network interface.

- We then created a network security group to secure our virtual machine.

- Once we had the basics, we launched a Windows virtual machine, attaching a public IP address directly to the network interface.

- Finally, we deployed a virtual machine extension that executed the PowerShell script on our Windows host to enable the WinRM protocol, allowing us to connect to and interact with the host using Ansible.

This chapter will repeat, tweak, and add to these steps using Ansible and the Azure collection of modules.

## Preparing Ansible for Microsoft Azure

Before we dive into the Ansible role, which will launch our resources, we need to do a little preparation; first, let's ensure that the Azure collection is installed by running the following:

```
$ ansible-galaxy collection install azure.azcollection
```

Next, we must install the Python modules that allow the Azure collection to interact with the Azure APIs. To do this, we need to run the following command:

```
$ pip3 install -r ~/.ansible/collections/ansible_collections/azure/
azcollection/requirements-azure.txt
```

With the necessary supporting Python modules installed, the next step is to ensure that you have signed into your Microsoft Azure account using the Azure CLI. To do this, run the following command and follow the onscreen prompts:

```
$ az login
```

If you have access to more than one Azure subscription using your account, you should ensure that the subscription you intend to launch your resources in is selected.

To do this, you can list all the subscriptions and, if needed, switch to the right subscription by running the following commands:

```
$ az account list --output table
$ az account set --subscription <subscription_id>
```

Ensure you replace `<subscription_id>` with the correct subscription ID from the `az account list` command.

> **Note**
>
> Using the `az account set` command only applies to your current session; if you close your terminal window and reopen a new session, you must ensure you have changed subscriptions again.

## Reviewing the variables

There are several variables across the roles we will use to deploy our Azure resources and configure WordPress. The first ones we will look at can be found in the `group_vars/common.yml` file.

To start with, we have some **feature flags**, the first of which, `debug_output`, outputs the contents of the variables that are registered during the playbook run; setting this to `true` is helpful to pull back information on the Azure resources once they have launched during the development of the role.

The second feature flag is `generate_key`; if this is set to `true`, then a private and public key pair will be created by Ansible if one does not exist at `~/.ssh/id_rsa`.

The playbook will use the key in this location when launching a virtual machine, so one must exist, as without it, Ansible cannot connect to the newly launched virtual machine.

These two variables look like the following:

```
debug_output: false
genterate_key: false
```

Next, in `group_vars/common.yml`, we define some information about our app workload; this contains a mixture of details about the application and some of the Azure details like the Azure region the workload will be launched in (`location` and `location_short`) as well as the name that our WordPress site will be accessible on (`public_dns_name`):

```
app:
   name: "learnansible-wordpress"
   shortname: "ansiblewp"
   location: "westeurope"
   location_short: "euw"
   env: "prod"
   public_dns_name: "learnansible"
```

The final set of variables, which are defined in the `group_vars/common.yml` file, are for the tags that will be applied to each Azure resource that Ansible will launch:

```
common_tags:
   "project": "{{ app.name }}"
   "environment": "{{ app.env }}"
   "deployed_by": "ansible"
```

The next set of variables we will be using can be found in `roles/azure/defaults/main.yml`, which are used to deploy our resources.

The first block of variables defines a quick dictionary of Azure service names for use when it comes to naming our resources:

```
dict:
  ansible_warning: "Resource managed by Ansible"
  load_balancer: "lb"
  network_interface: "nic"
  nsg: "nsg"
  private_endpoint: "pe"
  public_ip: "pip"
  resource_group: "rg"
  subnet: "snet"
  virtual_machine: "vm"
  virtualnetwork: "vnet"
```

Next, we define the resource names – as per *Chapter 7, Ansible Windows Modules*, I am naming the resources as close to the cloud adoption framework recommendations as possible:

```
load_balancer_name: "{{ dict.load_balancer }}-{{ app.name }}-{{app.
env}}-{{ app.location_short }}"
load_balancer_public_ip_name: "{{ dict.public_ip }}-{{ load_balancer_
name }}"
nsg_name: "{{ dict.nsg }}-{{ app.name }}-{{app.env}}-{{ app.location_
short }}"
resource_group_name: "{{ dict.resource_group }}-{{ app.name }}-{{app.
env}}-{{ app.location_short }}"
virtual_network_name: "{{ dict.virtualnetwork }}-{{ app.name }}-{{app.
env}}-{{ app.location_short }}"
vm_name: "{{ dict.virtual_machine }}-admin-{{ app.name }}-{{app.env}}-
{{ app.location_short }}"
vnet_name: "{{ dict.virtualnetwork }}-{{ app.name }}-{{app.env}}-{{
app.location_short }}"
```

Now that all the naming is out of the way, we can start defining the networking variables:

```
vnet_config:
  cidr_block: "10.0.0.0/24"
  subnets:
    - {
        name: "{{ dict.subnet }}-vms-{{ app.name }}-{{app.env}}-{{
app.location_short }}",
        subnet: «10.0.0.0/27»,
```

```
        private: true,
        service_endpoints: «Microsoft.Storage»,
    }
```

Next, in networking, we have two lists of IPs – one is for fixed IPs, and the other is the IP address discovered when the playbook runs:

```
trusted_ips:
  - ""
dynamic_ips:
  - "{{ your_public_ip }}"
```

The next block of variables takes the preceding lists of IP addresses and uses them when creating the two network security group rules:

```
nsg_rules:
  - name: "allowHTTP"
    description: "{{ dict.ansible_warning }}"
    protocol: «Tcp»
    destination_port_range: «80»
    source_address_prefix: "*"
    access: "Allow"
    priority: "100"
    direction: "Inbound"
  - name: "allowSSH"
    description: "{{ dict.ansible_warning }}"
    protocol: "Tcp"
    destination_port_range: «{{ load_balancer.ssh_port }}»
    source_address_prefix: "{{ trusted_ips|select() + dynamic_ips |
unique }}"
    access: "Allow"
    priority: "150"
    direction: "Inbound"
```

As you can see, the first rule, allowHTTP, opens port 80 to the world; but allowSSH locks down the SSH port to the IP addresses in our two lists. To do this, we take the list of IP addresses in the trusted_ips variable, append the content of dynamic_ips, and then finally only display the unique items in the list so any duplicate entries are removed.

The final networking block defines the basics needed to launch Azure Load Balancer:

```
load_balancer:
  ssh_port: "22"
  ssh_port_backend: "22"
```

```
http_port: "80"
http_port_backend: "80"
```

Now we have the virtual machine configuration:

```
vm_config:
  admin_username: "adminuser"
  ssh_password_enabled: false
  vm_size: "Standard_B1ms"
  image:
    publisher: "Canonical"
    offer: "0001-com-ubuntu-server-jammy"
    sku: "22_04-LTS"
    version: "latest"
  disk:
    managed_disk_type: "Premium_LRS"
    caching: "ReadWrite"
  key:
    path: "/home/adminuser/.ssh/authorized_keys"
    data: "{{ lookup('file', '~/.ssh/id_rsa.pub') }}"
```

Then, finally, just two variables that define the location and the host group of our newly launched virtual machine will be placed:

```
location: "{{ app.location }}"
hosts_group: "vmgroup"
```

Now that we have covered all of the variables needed to launch our Azure resources, we can work through the tasks that do the actual work, all of which can be found in `roles/azure/tasks/main.yml`.

## The resource group task

The first task we are going to look at is creating the resource group where all of the other Azure resources are going to be placed:

```
- name: "Create the resource group"
  azure.azcollection.azure_rm_resourcegroup:
    name: "{{ resource_group_name }}"
    location: "{{ location }}"
    tags: "{{ common_tags }}"
  register: "resource_group_output"
```

As you can see, there is not much to it; it takes the name, location, and tags variables we defined and creates the resource group using the azure. collection.azure_rm_resourcegroup module. The task output is then registered as a variable, allowing us to reuse the output in later tasks.

The next task prints the contents of the resource_group_output register variable on the screen if debug_output is set to true; if it is false, then the task is skipped:

```
- name: "Debug - Resource Group result"
  ansible.builtin.debug:
    var: "resource_group_output"
  when: debug_output
```

This is a common pattern throughout the Azure role, so we will not cover this task again. Assume that if the task registers its output, there is a supporting debug task immediately after. Now that we have our resource group, we can make a start on configuring the networking.

## The networking tasks

The first task launches the virtual network, placing it in the resource group we just created:

```
- name: "Create the virtual network"
  azure.azcollection.azure_rm_virtualnetwork:
    resource_group: "{{ resource_group_output.state.name }}"
    name: "{{ virtual_network_name }}"
    address_prefixes: "{{ vnet_config.cidr_block }}"
    tags: "{{ common_tags }}"
  register: "virtual_network_output"
```

As you can see, when referencing the resource group name, we use the registered output from the previous task by using {{ resource_group_output.state.name }}. Again, this is going to be a common thread throughout the remaining tasks.

Note, we are not defining the subnet as part of creating the virtual network; this is possible as we are only adding a single subnet, but it is considered best practice to use the azure.collection. azure_rm_subnet module to add subnets as this approach means that you can loop through adding subnets with a with_items statement:

```
- name: "Add the subnets to the virtual network"
  azure.azcollection.azure_rm_subnet:
    resource_group: "{{ resource_group_output.state.name }}"
    name: "{{ item.name }}"
    address_prefix: "{{ item.subnet }}"
    virtual_network: "{{ virtual_network_output.state.name }}"
    service_endpoints:
      - service: "{{ item.service_endpoints }}"
```

```
  with_items: "{{ vnet_config.subnets }}"
  register: "subnet_output"
```

With the virtual network now populated with subnets, we can move on to creating the network security group.

As you may remember, when we looked at the variables, we used a variable called `your_public_ip`, so our next task is to discover the external IP address of the host running Ansible using the `community.general.ipify_facts` module:

```
- name: "Find out your current public IP address using https://ipify.
org/"
  community.general.ipify_facts:
  register: public_ip_output
```

As you can see, there is not much to this, but we are not registering a variable called `your_public_ip`; this is done as a separate task that uses the `ansible.builtin.set_fact` module:

```
- name: "Register your public ip as a fact"
  ansible.builtin.set_fact:
    your_public_ip: "{{ public_ip_output.ansible_facts.ipify_public_ip
}}"
```

Now we know the IP address, we can create the network security group:

```
- name: "Create the network security group"
  azure.azcollection.azure_rm_securitygroup:
    resource_group: "{{ resource_group_output.state.name }}"
    name: "{{ nsg_name }}"
    rules: "{{ nsg_rules }}"
    tags: "{{ common_tags }}"
  register: "nsg_output"
```

So far, so good; the next piece of networking configuration we need to do is to launch Azure Load Balancer. This is the first deviation from the resources we launched in *Chapter 7, Ansible Windows Modules*, so why is that?

While Microsoft allows you to directly assign a public IP address to a virtual machine's network interface in Azure, it is generally frowned upon and not considered best practice – having a networking resource such as Azure Load Balancer to route and distribute your traffic to one or more hosts is deemed to be more secure as you are putting a layer between the virtual machine and the public internet.

Also, having traffic pass through a load balancer, even when running a single virtual machine like ours, allows you to perform a basic health check to see whether the port to which the load balancer sends traffic is healthy.

The first task we need to run when launching Azure Load Balancer is to create a public IP address resource, which will be attached to the load balancer when we launch it:

```
- name: "Create the public IP address needed for the load balancer"
  azure.azcollection.azure_rm_publicipaddress:
    resource_group: "{{ resource_group_output.state.name }}"
    allocation_method: "Static"
    name: "{{ load_balancer_public_ip_name }}"
    sku: "standard"
    domain_name: "{{ app.public_dns_name }}"
    tags: "{{ common_tags }}"
  register: "public_ip_output"
```

Now that the public IP address is defined, we can move on to Azure Load Balancer itself.

As there is rather a lot to the task, I will break it up a little as we go along:

```
- name: "Create load balancer using the public IP we created"
  azure.azcollection.azure_rm_loadbalancer:
    resource_group: "{{ resource_group_output.state.name }}"
    name: "{{ load_balancer_name }}"
    sku: "Standard"
```

The next block in the task defines the front of the load balancer. This is where we attach the public IP address we just created:

```
    frontend_ip_configurations:
      - name: "{{ load_balancer_name }}-frontend-ip-config"
        public_ip_address: "{{ public_ip_output.state.name }}"
```

Next up, we define the backend pool. This is the pool where our virtual machine will be placed to have traffic sent to it. If we had more than one virtual machine, all of them would be addressed to the pool:

```
    backend_address_pools:
      - name: "{{ load_balancer_name }}-backend-address-pool"
```

Now we have the health probe, which probes the HTTP port on the backend pool to make sure that the virtual machines are ready to access traffic on port 80 by seeing if the port is open:

```
    probes:
      - name: "{{ load_balancer_name }}-http-probe"
        port: «{{ load_balancer.http_port_backend }}»
        fail_count: "3"
        protocol: "Tcp"
```

For our WordPress workload, we want our HTTP port to be exposed. To do this, we will create a load-balancing rule that allows you to create a one-to-many relationship with one or more virtual machines in the backend pool. This rule exposes the HTTP port on the load balancer and sends the traffic to the HTTP port on the backend virtual machines. If we had more than one virtual machine, the traffic would be evenly distributed across all hosts in the backend on the HTTP port:

```
load_balancing_rules:
  - name: "{{ load_balancer_name }}-rule-http"
    frontend_ip_configuration: "{{ load_balancer_name }}-frontend-
ip-config"
    backend_address_pool: "{{ load_balancer_name }}-backend-
address-pool"
    frontend_port: «{{ load_balancer.http_port }}»
    backend_port: "{{ load_balancer.http_port_backend }}"
    probe: "{{ load_balancer_name }}-http-probe"
```

While a load balancing rule takes traffic from a single port on the frontend and distributes it across multiple virtual machines in the backend pool, an inbound **NAT** (which stands for **Network Address Translation**) rule distributes traffic on a one-to-one basis, which makes it perfect for services such as SSH that are not meant to be distributed across multiple hosts:

```
inbound_nat_rules:
  - name: "{{ load_balancer_name }}-nat-ssh"
    frontend_ip_configuration: "{{ load_balancer_name }}-frontend-
ip-config"
    backend_port: "{{ load_balancer.ssh_port }}"
    frontend_port: "{{ load_balancer.ssh_port }}"
    protocol: "Tcp"
```

If we were to have more than one machine, we would add more rules that take different ports and map them to port 22 on the backend virtual machines. Typically, I would use high ports such as 2220 > 2229 so I don't clash with over services – 2220 would send traffic to port 22 on the first machine and 2221 would do the same for the second machine, and so on.

However, in this example, we just have a single host, so I am mapping port 22 to port 22.

Lastly, we will tag the resource and register the output:

```
    tags: "{{ common_tags }}"
  register: "load_balancer_output"
```

Now we have the load balancer, we need to create a network interface, which will be placed in the backend pool and attached to our virtual machine.

For those of you who have already looked at the Ansible Azure collection, you may have noticed a module called `azure.azcollection.azure_rm_networkinterface`, which is used to manage network interfaces. Hence, you'd assume that the task we will be looking at uses that. Well, you would be wrong.

While the pre-written module has pretty good feature parity with the API endpoint it interacts with, it is missing one key piece of functionality we require for our deployment: the ability to assign the network interface to a NAT rule.

However, all is not lost, and there is a workaround.

There is an Azure module whose only purpose is to interact with the Azure Resource Manager API directly, called `azure.collection.azure_rm_resource`, and by using this module, we can make an API call directly to the `Microsoft.Network/networkInterfaces` endpoint from within Ansible.

Having the ability to do this for any of the Azure Resource Manager APIs is quite powerful as it opens new features as soon as Microsoft releases them, and it means you don't have to wait for the Ansible Azure collection developers to write, test, and release the module.

It does come with one downside, though: using this method does add an additional layer of complexity to your playbook.

The following URL is the link to the REST API documentation, which covers the creation of a network interface: `https://learn.microsoft.com/en-us/rest/api/virtualnetwork/network-interfaces/create-or-update?view=rest-virtualnetwork-2023-05-01&tabs=HTTP`.

As we will see from working through the task, the general gist of what we are doing is constructing the URL of the API we would like to target and then constructing the request body detailed in the REST documentation.

To start with, let's look at the part of the task that generates the URL:

```
- name: "Create the network interface for the wordpress vm"
  azure.azcollection.azure_rm_resource:
    api_version: "2023-05-01"
    resource_group: "{{ resource_group_output.state.name }}"
    provider: "network"
    resource_type: "networkinterfaces"
    resource_name: "{{ dict.network_interface }}-{{ vm_name }}"
    idempotency: true
```

The preceding information constructs the URL given in the documentation, which is the following:

```
PUT https://management.azure.com/subscriptions/{subscriptionId}/
resourceGroups/{resourceGroupName}/providers/Microsoft.Network/
networkInterfaces/{networkInterfaceName}?api-version=2023-05-01
```

Let us look at how this is generated:

- `{subscriptionId}` is automatically generated by the module, and we do not need to provide this information.

- `{resourceGroupName}` is added by providing the `resource_group` key, and as per the rest of the tasks, we are using the resource group name, which is the output of our registering the variable in the resource group task.

- The providers are provided by us by filling in the `provider` and `resource_type` keys. Don't worry – the URL is not case-sensitive, and the module adds the `Microsoft.` part for us.

- `{networkInterfaceName}` is the `resource_name` key.

- Finally, the API version is provided by filling in the `api_version` key.

The last part of the "header" does not form part of the URL, but instead, it instructs Ansible to perform a GET request and then compares the body of what will be posted to what is returned by the GET request, and if there are any problems, it will error before the body is posted.

Now that we have the URL of the Azure Resource Manager API endpoint to which we will send our request, we need to populate the body of the request.

For our case, this looks like the following code:

```yaml
body:
  location: "{{ location }}"
  properties:
    enableAcceleratedNetworking: false
    primary: true
    networksecuritygroup:
      id: "{{ nsg_output.state.id }}"
    configurations:
      - name: "{{ vm_name }}-ipcfg"
        properties:
          subnet:
            id: "{{ subnet_output.results[0].state.id }}"
          loadBalancerBackendAddressPools:
            - id: "{{ load_balancer_output.state.backend_address_
pools[0].id }}"
          loadBalancerInboundNatRules:
            - id: "{{ load_balancer_output.state.inbound_nat_
rules[0].id }}"
    tags: "{{ common_tags }}"
```

When the module runs, `properties` will be rendered as JSON and posted alongside `location` and `tags` in the request's body, leaving the final part of the task to register the output:

```
register: "network_interface_output"
```

We now have all the base Azure configuration and resources in place; we can launch our virtual machine. As we will be using SSH to connect to the virtual machine and bootstrap our WordPress installation, we need to ensure we have a valid SSH key generated.

As we will connect to a remote virtual machine, we want to ship a test key as we have been doing on our locally deployed hosts. So, if there is not a key at `~/.ssh/id_rsa` on your local machine, then set the `genterate_key` variable in the `group_vars/common.yml` file to `true` (it is `false` by default), then Ansible will generate the key for you.

Do not worry if a key already exists at that location; Ansible will only create a key if one does not exist:

```
- name: "Check user has a key, if not create one for {{ ansible_user_
  id }}"
  ansible.builtin.user:
    name: "{{ ansible_user_id }}"
    generate_ssh_key: true
    ssh_key_file: "~/.ssh/id_rsa"
  when: genterate_key
```

Next, we have the task that launches the virtual machine itself. It uses all of the resources we have already deployed and configured so I will not go into too much detail:

```
- name: Create the admin virtual machine
  azure.azcollection.azure_rm_virtualmachine:
    resource_group: "{{ resource_group_output.state.name }}"
    name: "{{ vm_name }}"
    admin_username: "{{ vm_config.admin_username }}"
    ssh_public_keys:
      - path: "{{ vm_config.key.path }}"
        key_data: "{{ vm_config.key.data }}"
    ssh_password_enabled: "{{ vm_config.ssh_password_enabled }}"
    vm_size: "{{ vm_config.vm_size }}"
    managed_disk_type: "{{ vm_config.disk.managed_disk_type }}"
    network_interfaces: "{{ network_interface_output.response.name }}"
    image:
      offer: "{{ vm_config.image.offer }}"
      publisher: "{{ vm_config.image.publisher }}"
      sku: "{{ vm_config.image.sku }}"
```

```
      version: «{{ vm_config.image.version }}»
    tags: «{{ common_tags }}»
  register: «vm_output»
```

As with most tasks we have run in this role, immediately after there is a debug task.

You may think, *"That's the end of the role, right?"* but we have two tasks to cover.

The first of these final two tasks takes information about the hosts, such as the public IP address and SSH port, and then adds it to the host group defined as the hosts_group variable.

This means that there is no hardcoding of IP addresses or connections in our host's inventory file. The task to register the host looks like the following:

```
- name: Add the Virtual Machine to the host group
  ansible.builtin.add_host:
    groups: "{{ hosts_group }}"
    hostname: "{{ public_ip_output.state.ip_address }}-{{ load_
balancer.ssh_port }}"
    ansible_host: "{{ public_ip_output.state.ip_address }}"
    ansible_port: «{{ load_balancer.ssh_port }}»
```

So, what could this task be? We have the networking in place, our virtual machine has been launched, and we have registered our host, so we must be ready to start bootstrapping WordPress.

That's the problem; we might be ready, but the host we just launched might not be as it can sometimes take a minute or two for the virtual machine to finish booting up. If we were to immediately try and SSH into the host before it has finished booting, then our playbook would error and halt running.

Luckily, an Ansible module was developed for use in this scenario, ansible.builtin.wait_for:

```
- name: "Wait for the virtual machine to be ready"
  ansible.builtin.wait_for:
    host: "{{ public_ip_output.state.ip_address }}"
    port: «{{ load_balancer.ssh_port }}»
    delay: 10
    timeout: 300
```

This will wait for 10 seconds and then attempt to SSH to the host for up to 5 minutes (300 seconds); when SSH is accessible, the Ansible playbook will then progress to the next set of roles, which, in our case, bootstrap WordPress.

# Bootstrapping WordPress

It won't be of any surprise to you that the bulk of the WordPress roles remain intact from our previous chapters so we will not cover those parts there and will instead review some of the small changes.

## The site and host environment files

The `site.yml` is now split into two sections; the first runs locally and interacts with the Azure Resource Manager API to launch and configure the Azure resources:

```
- name: "Deploy and configure the Azure Environment"
  hosts: localhost
  connection: local
  gather_facts: true
  vars_files:
    - group_vars/common.yml

  roles:
    - "azure"
```

The second section targets the `vmgroup` host group and looks more like what we have been working with so far in the previous chapters:

```
- name: "Install and configure Wordpress"
  hosts: vmgroup
  gather_facts: true
  become: true
  become_method: "ansible.builtin.sudo"
  vars_files:
    - group_vars/common.yml

  roles:
    - "secrets"
    - "stack_install"
    - "stack_config"
    - "wordpress"
```

The `hosts` file looks like the `hosts` files we have been using throughout the previous chapters; it is just missing the lines where we explicitly define the target hosts and instead is just made up of the host groups' definitions.

You may have noticed that we are adding a new role, and the remaining ones are mostly the same; the role is called `secrets`, so let's see what it does.

## The secrets role

The sole purpose of this role is to generate secure passwords for WordPress and the database. Its tasks are delegated to the local machine as it creates a variables file at `group_vars/secrets.yml` and loads them into the playbook run.

First, it checks if `group_vars/secrets.yml` already exists and if it does, we don't want to change the contents of the file:

```
- name: "Check if the file secrets.yml exists"
  ansible.builtin.stat:
    path: "group_vars/secrets.yml"
  register: secrets_file
  delegate_to: "localhost"
  become: false
```

If there is no file, then it and its contents are generated from a template file:

```
- name: "Generate the secrets.yml file using a template file if not
exists"
  ansible.builtin.template:
    src: "secrets.yml.j2"
    dest: "group_vars/secrets.yml"
  when: secrets_file.stat.exists == false
  delegate_to: "localhost"
  become: false
```

The template file at `roles/secrets/templates/secrets.yml.j2` looks like the following:

```
db_password: "{{ lookup('community.general.random_string', length=20,
upper=true, special=false, numbers=true) }}" wp_password: "{{
lookup('community.general.random_string', length=20, upper=true,
special=true, override_special="@-&*", min_special=2, numbers=true)
}}"
```

As you can see, it uses the `community.general.random_string` module to generate a random string with some sensible rules, which we will use as passwords.

## Other changes

Most of the changes to the roles are to the variables; for example, in `roles/wordpress/defaults/main.yml` we have the following:

```
wordpress:
  domain: "http://{{ app.public_dns_name }}.{{ app.location
}}.cloudapp.azure.com/"
  password: "{{ wp_password }}"
```

This uses the public URL we are configuring on the Azure Load Balancer public IP address and the password variable from the `secrets` role that just ran.

Everything else in the roles remains as we left it in *Chapter 5, Deploying WordPress*.

# Running the playbook

Running the playbook uses the same command we have been running throughout the book:

```
$ ansible-playbook -i hosts site.yml
```

The playbook will execute and by the end of it you should see something like the output on the following screen:

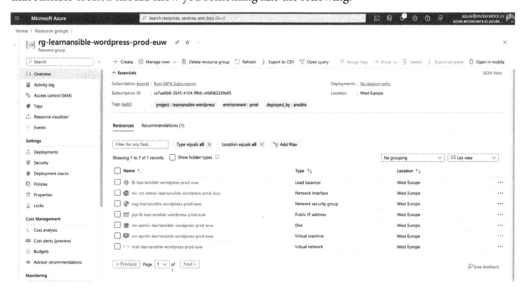

Figure 9.1 – Running the playbook in a terminal

Visiting the Azure portal at `https://portal.azure.com/` and viewing the resource group that Ansible created should show you something like the following:

Figure 9.2 – Viewing the resources in the Azure portal

From here, you should be able to enter the DNS name assigned on the public IP address; for example, in my instance, it was `http://learnansible.westeurope.cloudapp.azure.com/`. This may be different in your case and you should see your newly bootstrapped WordPress site.

Just like when we launched Azure resources in *Chapter 7, Ansible Windows Modules*, to terminate the resources, we need to remove the resource group, which will remove all the resources contained there.

To do this using Ansible, there is a small, self-contained playbook called `destroy.yml`, which can be executed by running the following:

```
$ ansible-playbook -i hosts destory.yml
```

This will take a few minutes to run, but it will remove all resources deployed in the *site.yml* playbook, including the ones in Azure and the `group_vars/secrets.yml` file, leaving you with a nice clean slate for when you next run the main `site.yml` playbook.

## Summary

In this chapter, we launched our first instances in a public cloud using the Azure Ansible modules; as you have seen, the process was relatively straightforward, and we managed to securely launch the network and compute resource in Microsoft Azure, ready for us to then install WordPress on it without making any significant changes to the roles we covered in *Chapter 5, Deploying WordPress*.

In the next chapter, we will expand on some of the techniques we have covered in this chapter and return to networking, but unlike the previous chapter, where we covered networking devices, we will be looking at networking in public clouds.

# 10

# Building Out a Cloud Network

Now that we have launched servers in Microsoft Azure, we will start looking at launching services within **Amazon Web Services** (**AWS**).

Before we launch virtual machine instances, we must create a network to host them. This is called a **virtual private cloud** (**VPC**) and there are a few different elements we will need to bring together in a playbook to create one, which we will then be able to use for our instances.

In this chapter, we will do the following:

- Receive an introduction to AWS
- Cover what it is we are trying to achieve and why
- Create a VPC, subnets, and routes (networking and routing)
- Create security groups (firewall)

We will look at more advanced Ansible techniques as we launch and manage more dynamic resources with complex dependencies.

The chapter covers the following topics:

- An introduction to AWS
- Amazon VPC overview
- Creating an access key and secret
- Getting Ansible ready for targeting AWS
- The AWS playbook
- Running the playbook

# Technical requirements

This chapter will use AWS; you will need administrator access to create the roles needed to allow Ansible to interact with your account. As with other chapters, you can find the complete playbooks in the `Chapter10` folder in the accompanying GitHub repository at `https://github.com/PacktPublishing/Learn-Ansible-Second-Edition/tree/main/Chapter10`.

# An introduction to AWS

AWS has been around since 2002; it started by offering a few services that were not linked in any way. It progressed in this form until early 2006 when it was relaunched. The relaunched AWS brought together three services:

- **Amazon Elastic Compute Cloud** (**Amazon EC2**): This is the AWS compute service
- **Amazon Simple Storage Service** (**Amazon S3**): Amazon's scalable object storage service
- **Amazon Simple Queue Service** (**Amazon SQS**): This service provides a message queue, primarily for web applications

Since 2006, it has grown from three unique services to over 160, covering over 15 primary areas such as the following:

- Compute
- Storage
- Database
- Networking and content delivery
- Machine learning analytics security, identity, and compliance
- Internet of things

At its earnings call in October 2023, it was revealed that AWS had USD 23.06 billion in revenue in the third quarter of 2023, good for a service that initially offered to share idle compute time.

At the time of writing, AWS spans 32 geographic regions, which host a total of 102 availability zones (`https://aws.amazon.com/about-aws/global-infrastructure/`).

So, what makes AWS so successful? Not only its coverage but its approach to putting out its services. Andy Jassy, AWS CEO, has been quoted as saying:

> *"Our mission is to enable any developer or any company to be able to build all their technology applications on top of our infrastructure technology platform."*

As an individual, you have access to the same APIs, services, regions, tools, and pricing models as large multi-national companies and Amazon themselves, as they consume their services. This gives you the freedom to start small and scale massively. For example, Amazon EC2 instances start from around USD 4.50 per month for a t2.nano (1 vCPU and 0.5G) all the way up to over USD 19,000 per month for an x1e.32xlarge (128 vCPU, 3,904 GB RAM, and two 1920 GB SSD storage); as you can see, there are instance types for every workload imaginable.

Both instances and most services are billed under pay-as-you-go, from per-second billing for EC2 instances to pay per GB per month for the storage you are using.

## Amazon VPC overview

In this chapter, we are going to be concentrating on launching an **Amazon Virtual Private Cloud** (**Amazon VPC**); this is the networking layer that will host the computing and other Amazon services that we will be launching in *Chapter 11*, *Highly Available Cloud Deployments*.

We are going to be launching our VPC into the **EU-West #1 (Ireland)** region; we will be spanning all three availability zones for our **EC2** instances and also the **Application Elastic Load Balancer**. We will, again, be using the three availability zones for our **Amazon Relational Database Service** (**RDS**) instance and also two zones for the **Amazon Elastic File System** (**Amazon EFS**) volumes.

This all means our Ansible playbook needs to create/configure the following:

- One Amazon VPC
- Three subnets for EC2 instances
- Three subnets for Amazon RDS instances
- Three subnets for Amazon EFS volumes
- Three subnets for the Application Load Balancer
- One internet gateway

We will also need to configure the following:

- One route to allow access through the internet gateway
- One security group that allows everyone to access port 80 (HTTP) and 443 (HTTPS) on the Application Load Balancer
- One security group that allows trusted source access to port 22 (SSH) on the EC2 instances
- One security group that allows access to port 80 (HTTP) from the Application Load Balancer to the EC2 instances

- One security group that allows access to port 3306 (MySQL) on the Amazon RDS instances from the EC2 instances

- One security group that allows access to port 2049 (NFS) on the Amazon EFS volumes from the EC2 instances

This will give us our primary network, allowing restrictive access to everything but the Application Load Balancer, which we want to be publicly available.

Before creating an Ansible playbook that deploys the network, we need to get an AWS API access key and secret.

## Creating an access key and secret

It is more than possible to create an access key and secret key for your AWS user to give Ansible full access to your AWS account.

Because of this, we are going to look at creating a user for Ansible, which only has permission to access the parts of AWS we know that Ansible will need to interact with for the tasks we are covering in this chapter. We will be giving Ansible full access to the following services:

- Amazon VPC

- Amazon EC2

- Amazon RDS

- Amazon EFS

To do this, follow these steps:

1. Log in to the AWS console, which can be found at https://console.aws.amazon.com/.

2. Once logged in, click on **Services**, which can be found in the menu at the top of the screen. In the open menu, enter IAM into the search box and then click on the **IAM "Manage access to AWS resources"** result.

3. On the **IAM** page, click **User Groups** in the left-hand side menu; we will create a group with the permissions assigned to it, and then we will create a user and add it to our group.

4. Once on the **User Groups** page, click the **Create Group** button. This process has two steps, the first of which is setting the group's name. In the space provided, enter the group name Ansible.

5. Now, in the **Attach permissions policies – Optional** section, select **AmazonEC2FullAccess**, **AmazonVPCFullAccess**, **AmazonRDSFullAccess**, and **AmazonElasticFileSystemFullAccess**; once all four have been selected, click on the **Create Group** button at the bottom of the page.

6. Now that we have our Ansible group, click **Users** in the left-hand side menu.

7. Once on the **Users** page, click **Create user**, and this will take you to a page where you can configure your desired username and the type of user you want. Enter the following information:

   - **User name**: Enter `LearnAnsible` in here

   - Leave the **Provide user access to the AWS Management Console – optional** option unchecked, as we will create a programmatic user

8. Click on the **Next** button to take you to the **Set Permissions** page. Ensure that **Add user to group** is selected and that you have the `Ansible` group we created earlier ticked, and then click **Next**, which will take you to the **Review and Create** page.

9. Once you have reviewed the details, you need to click the **Create user** button, which will precisely do that: create our `LearnAnsible` user.

10. The final step is to get an access key for our user. To get this, click on the `LearnAnsible` user and select the **Security credentials** tab; from there, scroll down to **Access Keys** and click the **Create access key** button.

11. In the list of **Access key best practices & alternatives** select **Other** and then the **Next** button. Enter `For use with Learn Ansible` for the description tag value and then click **Create access key**.

12. The **Retrieve access keys** page is the only time you get access to the Secret access key, so I recommend downloading the CSV file. Once downloaded, click on **Done**.

---

Important note

The CSV file you have just downloaded contains credentials allowing whoever has them to launch resources in your AWS account; please do not share them and keep them safe, as they could be misused, resulting in a huge and unexpected AWS bill should they fall into the wrong hands.

---

Now that we have an access key ID and secret access key for a user with the permissions, we need to launch our VPC using Ansible; we can start getting Ansible ready and reviewing the playbook.

## Getting Ansible ready for targeting AWS

We first need to discuss how to pass our access key ID and secret access key to Ansible safely and securely. As I will share the final playbooks in a public repository on GitHub, I want to keep my AWS keys private from the world as that could get expensive! Typically, if it were a private repository, I would use Ansible Vault or some other secret management to encrypt the keys and include them with other potentially sensitive data, such as deployment keys.

In this case, I don't want to include any encrypted information in the repository, as it would mean that people would need to unencrypt it, edit the values, and then re-encrypt it. Luckily, the AWS modules allow you to set two environment variables on your Ansible controller; those variables will then be read as part of the playbook execution.

To set the variables, run the following commands to make sure that you replace the content with your access key and secret after = (the information listed as follows is just placeholder values):

```
$ export AWS_ACCESS_KEY=AKIAI5KECPOTNTTVM3EDA
$ export AWS_SECRET_KEY=Y4B7FFiSW10Am3VIFc071gnc/TAtK5+RpxzIGTr
```

Once set, you can view the contents by running the following:

```
$ echo $AWS_ACCESS_KEY
```

Now that we can securely pass our credentials to Ansible, we can install the Python modules needed by the AWS Ansible modules to interact with the AWS API.

> **Important note**
>
> You must set the environment variables for each terminal session, as they will be lost each time you close your terminal.

To install the Python modules, run the following command:

```
$ pip3 install botocore boto3
```

Now that we have the basics configured, we can review our playbook.

## The AWS playbook

As mentioned at the start of the chapter, we are going to be using some more advanced techniques when it comes to deploying resources in AWS where possible; I have tried to allow the resources to be deployed as dynamically as possible, a lot of which comes down to how we define our variables, which is where we are going to start our playbook review.

### The playbook variables

Most of the variables we define can be found in group_vars/common.yml, and as you can see from the following, they start by looking a lot like the variables we described in *Chapter 9, Moving to the Cloud*:

```
debug_output: false
app:
  name: "learnansible"
  region: "eu-west-1"
  env: "prod"
```

As you can see, we have the same debug_output feature flag and selection of variables used to describe our app and the AWS region in which it will be launched.

Next up, we have the resource names:

```
vpc_name: "{{ app.name }}-{{ app.env }}-{{ playbook_dict.vpc }}"
internet_gateway_name: "{{ app.name }}-{{ app.env }}-{{ playbook_dict.
internet_gateway }}"
internet_gateway_route_name: "{{ internet_gateway_name }}-{{ playbook_
dict.route }}"
```

Nothing too out of the ordinary so far, but here we will find our first difference in approach:

```
vpc:
  cidr_block: "10.0.0.0/23"
  dns_hostnames: true
  dns_support: true
  subnet_size: "27"
  subnets:
    - name: "ec2"
      role: "{{ subnet_role_compute }}"
    - name: "rds"
      role: "{{ subnet_role_database }}"
    - name: "efs"
      role: "{{ subnet_role_storage }}"
    - name: "dmz"
      role: "{{ subnet_role_public }}"
```

At first glance, that doesn't look too dissimilar to what we did for Microsoft Azure.

However, you might have noticed that there are no IP address CIDR ranges listed for the subnets, just some details about the subnets, including a dictionary of roles:

```
subnet_role_compute: "compute"
subnet_role_database: "database"
subnet_role_storage: "storage"
subnet_role_public: "public"
```

We will look at why the subnet's CIDR ranges are missing when we get to the tasks that create the subnet.

Next, we have the variables for creating the security groups; in total, we will be configuring four security groups, so in the interest of space, I will only be showing one of the small groups here:

```
security_groups:
  - name: "{{ app.name }}-rds-{{ playbook_dict.security_group }}"
    description: "opens port 3306 to the ec2 instances"
    id_var_name: "rds_group_id"
    rules:
      - proto: "tcp"
```

```
        from_port: "3306"
        to_port: "3306"
        group_id: "{{ ec2_group_id | default('') }}"
        rule_desc: "allow {{ ec2_group_id | default('') }} access to
port 3306"
```

See the GitHub repo for the full configuration for the four security groups; there is only one thing at this point to highlight, and that is this: where we reference `{{ ec2_group_id | default('') }}`, we are setting a default value of nothing (which is the `''` part). We will discuss why we are doing this when we cover the security role.

The final set of variables is the dictionary (`playbook_dict`) and a variable, which sets the value of `region` using `app.region`; again, see the GitHub if you want to see all the contents.

## The VPC role

Before we get to the exciting tasks, we need to create the VPC. The task in `roles/vpc/tasks/main.yml` looks like the following:

```
- name: "Create VPC"
  amazon.aws.ec2_vpc_net:
    name: "{{ vpc_name }}"
    region: "{{ region }}"
    cidr_block: "{{ vpc.cidr_block }}"
    dns_hostnames: "{{ vpc.dns_hostnames }}"
    dns_support: "{{ vpc.dns_support }}"
    state: "{{ state }}"
    tags:
      Name: "{{ vpc_name }}"
      projectName: "{{ app.name }}"
      environment: "{{ app.env }}"
      deployedBy: "{{ playbook_dict.deployedBy }}"
      description: "{{ playbook_dict.ansible_warning }}"
  register: vpc_output
```

The task is pretty much as you would expect, apart from the tags being set a little more in line than those we defined in *Chapter 9, Moving to the Cloud*. There is also a debug statement that prints the results of creating the VPC if you set `debug_output` to `true`:

```
- name: "Debug - VPC result"
  ansible.builtin.debug:
    var: "vpc_output"
  when: debug_output
```

From now on, it is safe to assume that all registered output will be followed by an `ansible.builtin.debug` task. Now that we have our VPC launched, we can start putting things inside it, beginning with the subnets, where things get more interesting.

## The subnets role

As mentioned in the AWS overview, there are 32 geographic regions and, at the time of writing, 102 Availability Zones. AWS differs from Microsoft Azure in that you need a subnet per Availability Zone rather than a single subnet spanning all the availability zones.

The `eu-west-1` region, which is the region we will target, is made up of three availability zones, and we have subnets for four different roles, meaning that we need 12 subnets in total, but our playbook could easily be targeting a region that only has two availability zones, or in some cases, even more.

So, our first task is to get information on the availability zones in our target region:

```
- name: "Get some information on the available zones"
  amazon.aws.aws_az_info:
    region: "{{ region }}"
  register: zones_output
```

Now that we know some information on the region, we can use that information and create our subnets:

```
- name: "Create all subnets"
  ansible.builtin.include_tasks: create_subnet.yml
  loop: "{{ vpc.subnets }}"
  loop_control:
    loop_var: subnet_item
    index_var: subnet_index
  vars:
    subnet_name: "{{ subnet_item.name }}"
    subnet_role: "{{ subnet_item.role }}"
    az_zones_from_main: "{{ zones_output }}"
  register: subnet_output
```

This task is quite different from the ones we have been using so far in the book, so let's take a deeper dive into what is happening.

Here, we are using a loop to automate the creation of multiple subnets. Each iteration of the loop processes one subnet from the `vpc.subnets` list, which, as we have already seen, contains the configuration details for each subnet.

As the loop runs, it assigns the current subnet's details to the `subnet_item` variable and its index in the list to `subnet_index`. These variables are then utilized to customize the creation process for each subnet.

The task includes and executes the steps defined in `create_subnet.yml` (which we will cover next) for each subnet, using the specific details of that subnet (such as its name and role).

You may have noticed that we still haven't passed in any CIDR ranges for the subnets; this is all handled within the `create_subnet.yml` task, which we loop over for each of our four subnet types; this is also where a second loop happens:

```
- name: "Create subnet in the availability zone"
  amazon.aws.ec2_vpc_subnet:
    region: "{{ region }}"
    state: "{{ state }}"
    vpc_id: "{{ vpc_output.vpc.id }}"
    cidr: "{{ vpc_output.vpc.cidr_block | ansible.utils.ipsubnet(vpc.
subnet_size, az_loop_index + (subnet_index * az_zones_from_main.
availability_zones|length)) }}"
    az: "{{ az_item.zone_name }}"
    tags:
      Name: "{{ subnet_name }}-{{ playbook_dict.subnet }}-{{ az_item.
zone_id }}"
      projectName: "{{ app.name }}"
      environment: "{{ app.env }}"
      deployedBy: "{{ playbook_dict.deployedBy }}"
      description: "{{ playbook_dict.ansible_warning }}"
      role: "{{ subnet_role }}"
  loop: "{{ az_zones_from_main.availability_zones }}"
  loop_control:
    loop_var: az_item
    index_var: az_loop_index
```

Please stick with me, as this is where it gets a little confusing; for each of the four loops we are enacting from our main loop, we are taking the information on the availability zones and then looping over them, creating a subnet per availability zone for the role we are currently looping over.

So, what about the CIDR range for the subnet?

You may have noticed something where you would expect to see the CIDR range; we have this expression:

```
vpc_output.vpc.cidr_block | ansible.utils.ipsubnet(vpc.subnet_size,
az_loop_index + (subnet_index * az_zones_from_main.availability_
zones|length))
```

We have the following components in the expression:

- `vpc_output.vpc.cidr_block`: This is the CIDR block of the VPC, within which the subnets will be created. For our example, it's `10.0.0.0/22`.

- `vpc.subnet_size`: This specifies the size of each subnet. We are using `/27`, representing a subnet with 32 IP addresses.

- `az_zones_from_main.availability_zones|length`: This is the total number of availability zones available. The region we are targeting has 3 availability zones.
- `az_loop_index`: This is the current index in the loop over the availability zones.
- `subnet_index`: This is the index of the current subnet being processed.

This means that for our expression, we will get the following results. The first subnet, which is labeled **ec2**, in the availability zone (`az1`) will have the following:

- `az_loop_index = 0`
- `subnet_index = 0`

So, the formula would be $0+(0*3)=0$, meaning that we would get the following:

```
cidr = "{{ vpc_output.vpc.cidr_block | ansible.utils.ipsubnet(27, 0)
}}"
```

With `vpc_output.vpc.cidr_block` being $10.0.0.0/22$, we could get the first $/27$, which would be $10.0.0.0/27$.

For the second availability zone (`az2`), the loop would be the following:

- `az_loop_index = 1`
- `subnet_index = 0`

$1+(0*3)=1$ means we would get $10.0.0.32/27$ since the next subnet block starts immediately after the previous one at the next 32 IP address interval.

The third Availability Zone (`az3`) would be $2+(0*3)=2$, and the CIDR block would be $10.0.0.64/27$.

The next subnet role, which is the RDS role, would give the following for `az1`:

- `az_loop_index = 0`
- `subnet_index = 1`

The formula would be $0+(1*3)=3$, giving us a CIDR block $10.0.0.96/27$.

This pattern would follow the sequence, where the next subnet for RDS `az2` would be at $10.0.0.128/27$, and for `az3`, it would be at $10.0.0.160/27$, and so on.

This expression ensures that each subnet created within the VPC is assigned a unique and non-overlapping CIDR block, segmented adequately according to the defined subnet size, and distributed across different availability zones.

Taking this approach not only simplifies the management of subnet creation but also ensures efficiency when it comes to writing the role, as it means that we don't have to hardcode tasks to consider changes between regions or the number of subnets we are defining in our variables.

The remaining tasks in the role build a list of the subnet IDs for each of the roles we have defined. An example of one of these tasks is as follows:

```
- name: "Gather information about the compute subnets"
  amazon.aws.ec2_vpc_subnet_info:
    region: "{{ region }}"
    filters:
      "tag:role": "{{ subnet_role_compute }}"
      "tag:environment": "{{ app.env }}"
      "tag:projectName": "{{ app.name }}"
  register: subnets_compute_output
```

This gets information on the three subnets assigned the `subnet_role_compute` role. A few more of these data-gathering tasks can be found in the repo; these cover the `subnet_role_database`, `subnet_role_storage`, and `subnet_role_public` roles.

Finally, the final task in the role prints the subnet IDs that we have gathered using the previous set of tasks; this looks slightly different to the debug statements we have been using in the playbook so far, as we are using the `msg` function rather than the `var` one when calling the `ansible.builtin.debug` module.

## The gateway role

The gateway role is relatively simple compared to the previous one. In comparison, it deploys an internet gateway. Then, it creates a route to send all traffic destined for the internet (represented by using `0.0.0.0/0`, the CIDR notation for all network traffic) to our newly launched internet gateway.

The task that creates the internet gateway looks like the following:

```
- name: "Create an Internet Gateway"
  amazon.aws.ec2_vpc_igw:
    region: "{{ region }}"
    state: "{{ state }}"
    vpc_id: "{{ vpc_output.vpc.id }}"
    tags:
      "Name": "{{ internet_gateway_name }}"
      "projectName": "{{ app.name }}"
      "environment": "{{ app.env }}"
      "deployedBy": "{{ playbook_dict.deployedBy }}"
      "description": "{{ playbook_dict.ansible_warning }}"
      "role": "igw"
  register: internet_gateway_output
```

As per the rest of the tasks, a debug task follows this, and then the task that creates the route table, which is then associated with our newly created internet gateway and also the computing and public subnets that we defined and gathered the information for in the subnet's role:

```
- name: "Create a route table so the internet gateway can be used by
  the public subnets"
  amazon.aws.ec2_vpc_route_table:
    region: "{{ region }}"
    state: "{{ state }}"
    vpc_id: "{{ vpc_output.vpc.id }}"
    subnets: "{{ subnet_compute_ids + subnet_public_ids }}"
    routes:
      - dest: "0.0.0.0/0"
        gateway_id: "{{ internet_gateway_output.gateway_id }}"
    resource_tags:
      "Name": "{{ internet_gateway_route_name }}"
      "projectName": "{{ app.name }}"
      "environment": "{{ app.env }}"
      "deployedBy": "{{ playbook_dict.deployedBy }}"
      "description": "{{ playbook_dict.ansible_warning }}"
      "role": "route"
  register: internet_gateway_route_output
```

We then do a debug task that completes this role, and we then move on to the final role of the playbook: the security group's role.

## The security group's role

While this role, in my opinion, is not as complicated as the subnet's role, we have built a little more logic into the task than some of the more straightforward tasks in the book that we have run so far.

If you recall, earlier in the chapter, when we covered the variables being used by the playbook, we gave the following example of the security groups being deployed:

```
- proto: "tcp"
  from_port: "3306"
  to_port: "3306"
  group_id: "{{ ec2_group_id | default('') }}"
  rule_desc: "allow {{ ec2_group_id | default('') }} access to port
3306"
```

The preceding rule, as per `rule_desc`, opens up port 3306 for any devices that have the EC2 security group attached to them, which, as we will see in *Chapter 11, Highly Available Cloud Deployments*, will be the EC2 instances that will be running our workload.

You may think to yourself, *"Now that makes sense."* However, this is a little bit of a flaw in the logic we must work around. ec2_group_id is referencing a group ID, which, at the time we first run our playbook, doesn't exist. So, how can we create the groups and populate them with rules that reference groups that don't yet exist?

As we have already seen, looping over the resources defined in our variables is more efficient. It reduces the hard-coded logic at the role level, making the role more re-useable between projects and playbooks.

Before we look at the logic of creating the groups, we need to gather one bit of information: the public IP address of the resource running Ansible. To do this, we call the following task:

```
- name: "Find out your current public IP address using https://ipify.
  org/"
  community.general.ipify_facts:
  register: public_ip_output
```

Then we set a fact called your_public_ip, which we can reference in our rules where needed:

```
- name: "Set your public ip as a fact"
  ansible.builtin.set_fact:
    your_public_ip: "{{ public_ip_output.ansible_facts.ipify_public_ip
}}/32"
```

Now that we have that snippet of information, we can return to the question of how we can reference the IDs of resources that have yet to be launched.

To create the security groups, we will be using the amazon.aws.ec2_security_group module. The module has a flag called purge_rules, set to true by default; in this default state, when our playbook finds and needs to update an existing security group, it will drop all the rules in the group and then add just the ones defined in the playbook to maintain a consistent state.

While it is a valid use case, in our example, disabling this functionality by setting purge_rules to false will allow us to create some unpopulated security groups:

```
- name: "Create the base security groups"
  amazon.aws.ec2_security_group:
    region: "{{ region }}"
    state: "{{ state }}"
    vpc_id: "{{ vpc_output.vpc.id }}"
    name: "{{ item.name }}"
    description: "{{ item.description }}"
    purge_rules: false
    tags:
      "Name": "{{ item.name }}"
      "projectName": "{{ app.name }}"
      "environment": "{{ app.env }}"
```

```
        "deployedBy": "{{ playbook_dict.deployedBy }}"
        "role": "securitygroup"
    loop: "{{ security_groups }}"
    register: base_security_groups_output
```

This will loop through and create the base, unpopulated security groups if they don't exist, and if they do already exist, no changes will be made to them.

So, now that we have our groups created, or if they already exist, we have the information we need to dynamically define some facts based on the output of the previous tasks:

```
- name: "Set the fact for the security group ids"
  ansible.builtin.set_fact:
    "{{ item.id_var_name }}": "{{ base_security_groups_output.
results | selectattr('item.name', 'equalto', item.name) |
map(attribute='group_id') | first }}"
  loop: "{{ security_groups }}"
  when: base_security_groups_output.results | selectattr('item.name',
'equalto', item.name) | map(attribute='group_id') | list | length > 0
```

This task uses the `ansible.builtin.set_fact` module, allowing the creation or update of new variables during runtime. This task aims to extract the unique ID of each security group created in the first task and assign it to a specific variable name.

There are two expressions we use to do this. The first is the following:

```
"{{ item.id_var_name }}": "{{ base_security_groups_output.
results | selectattr('item.name', 'equalto', item.name) |
map(attribute='group_id') | first }}"
```

This is used to create the dynamic set of variables based on the loop created by the second expression. A breakdown of this first expression follows:

- `base_security_groups_output.results`: This refers to the list of results from the previous task that created the security groups. Each result in this list contains data about one of the security groups.

- `selectattr('item.name', 'equalto', item.name)`: The `selectattr` filter is used to search through the list of results. It looks for results where the name attribute of the item (each security group) is equal to the current `item.name` in the loop. In other words, it filters the results to find the specific security group we're currently interested in.

- `map(attribute='group_id')`: The map filter is then used to transform the filtered list of results. It extracts only the `group_id` attribute from each result, which is the ID of the security group.

- `first`: Since the previous step can still return a list (albeit with a single element), the `first` filter takes only the first element from this list, which should be the unique ID of the security group.

The result of this expression is the ID of the security group that matches the current item in the loop, and it's assigned to a variable named according to `item.id_var_name`.

The second expression, which is in the when condition, runs as part of the loop:

```
when: base_security_groups_output.results | selectattr('item.name',
'equalto', item.name) | map(attribute='group_id') | list | length > 0
```

This expression determines whether the task should be executed for a particular item in the loop. It follows a similar logic to the first expression:

- It starts with the same filtering process to find the security group that matches the current `item.name`.

- After extracting the `group_id`, it ensures the output is treated as a list using the `list` filter.

- `length > 0`: This part checks whether the length of the list (the number of items in it) is greater than 0. This means at least one security group with the specified name must exist. If the list is empty, no matching security group is found, and the task will be skipped for the current item.

In theory, we should have now populated the variables that contain the security group IDs, meaning that we can now add the rules:

```
- name: "Provision security group rules"
  amazon.aws.ec2_security_group:
    region: "{{ region }}"
    state: "{{ state }}"
    vpc_id: "{{ vpc_output.vpc.id }}"
    name: "{{ item.name }}"
    description: "{{ item.description }}"
    purge_rules: false
    rules: "{{ item.rules }}"
  loop: "{{ security_groups }}"
  register: security_groups_with_rules_output
```

This will loop over the already created groups and populate the rules for each one, using the group IDs from the variables we dynamically defined in the previous task.

## Running the playbook

As mentioned earlier, we worked our way through the playbook code; before you run the playbook, you must set the `AWS_ACCESS_KEY` and `AWS_SECRET_KEY` environment variables on your terminal session by running the following, making sure to update any values to those that you made a note of when you created the Ansible user in the AWS console:

```
$ export AWS_ACCESS_KEY=AKIAI5KECPOTNTTVM3EDA
$ export AWS_SECRET_KEY=Y4B7FFiSW10Am3VIFc071gnc/TAtK5+RpxzIGTr
```

With the environment variables set, you can run the playbook running the now very familiar following code:

```
$ ansible-playbook -i hosts site.yml
```

Once completed, you should see something like the following terminal output:

Figure 10.1 – Running the playbook in a terminal

Going to the VPC and viewing the resource map in `http://console.aws.amazon.com/` should display something like the following resource map:

Figure 10.2 – Viewing the resource map

By going to **Security Groups**, you should also see the groups that we created listed:

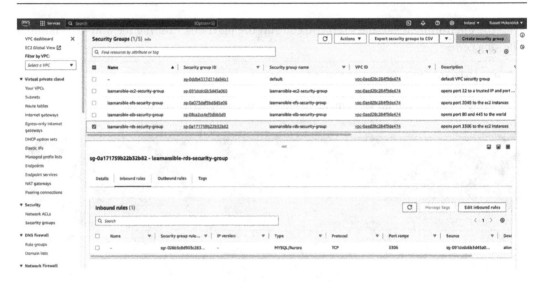

Figure 10.3 – Reviewing the security groups

I have included a second playbook in the repo, which destroys all of the resources created by running the `site.yml` playbook called `destroy.yml`. You can run it using the following command:

```
$ ansible-playbook -i hosts destroy.yml
```

I am not going to cover the contents of the playbook here, but if you review the code, you will notice that, in essence, it runs the same tasks in the role we have covered in this chapter in reverse order, setting the state to `absent` rather than `present`.

## Summary

In this chapter, we have taken our next step in using Ansible to launch resources in a public cloud. We have laid the groundwork for automating quite a complex environment by creating a VPC, setting up the subnets we need for our application, provisioning an internet gateway, and setting our instances to route their outgoing traffic through it.

We have configured four security groups, with three containing dynamic content, to secure the services launching into our VPC.

In the next chapter, we will build on the foundations laid in this chapter and launch a more complex set of services alongside the VPC.

## Further reading

- **Details of the AWS Q3 2023 earnings call**: https://www.cnbc.com/2023/10/26/aws-q3-earnings-report-2023.html

# 11
# Highly Available Cloud Deployments

Continuing with our AWS deployment, we will start to deploy services into the network we created in the previous chapter, and by the end of the chapter, we will be left with a highly available WordPress installation.

Building on top of the roles we created in the previous chapter, we will be doing the following:

- Launching and configuring an Application Load Balancer
- Launching and configuring Amazon **Relational Database Service (RDS)** (database)
- Launching and configuring Amazon **Elastic File System (EFS)** (shared storage)
- Launching an **Elastic Compute Cloud** (EC2) instance and creating an **Amazon Machine Image (AMI)** from it (deploying the WordPress code)
- Launching and configuring a launch template to use the newly created AMI and autoscaling group (high availability)

The chapter covers the following topics:

- Planning the deployment
- The Playbook
- Running the Playbook
- Terminating all the resources

# Technical requirements

As in the previous chapter, we will be using AWS; you will need the access key and secret key we created in the previous chapter to launch the resources needed for our highly available WordPress installation. Please note that we will be launching resources that incur charges. Again, you can find the complete playbook in the `Chapter11` folder of the accompanying GitHub repository at `https://github.com/PacktPublishing/Learn-Ansible-Second-Edition/tree/main/Chapter11/`.

# Planning the deployment

Before diving into the playbooks, we should get an idea of what we are trying to achieve. As mentioned, we will build on our AWS **Virtual Private Cloud** (**VPC**) role by adding instances and storage; our final deployment will look like the following diagram:

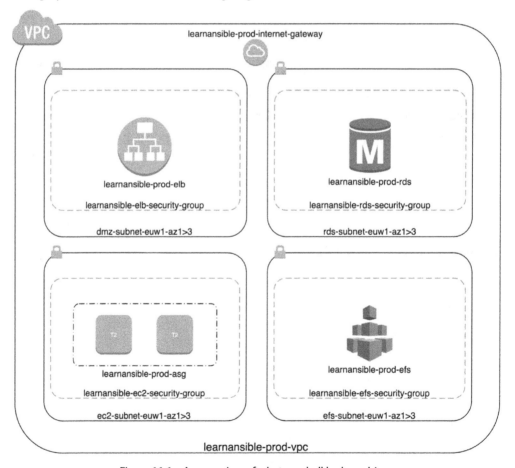

Figure 11.1 – An overview of what we shall be launching

In the diagram, we have the following:

- 2 x EC2 instances (t2.micro), deployed across different availability zones
- 1 x RDS instances (t2.micro)
- 1 x EFS storage across three availability zones

Before we talk about the deployment itself, based on the diagram and specifications here, how much is this deployment going to cost us to run?

## Costing the deployment

The cost of running this deployment in the EU-West-1 region is as follows:

| Instance Type | # Number | Instance cost | Total Monthly Cost |
|---|---|---|---|
| EC2 instances (t2.micro) | x2 | $9.20 | $18.40 |
| RDS instance (t2.micro) | x1 | $13.14 | $13.14 |
| Application Load Balancer | x1 | $24.24 | $24.24 |
| EFS | 5GB | $0.88 | $4.40 |
| Total | | | $61.83 |

Table 11.1 – Cost of running the deployment

There will be a few other minor costs, such as bandwidth and storing the AMI that contains our software stack. We could also consider increasing these costs by adding additional redundancy, such as updating our RDS instance to a multi-AZ RDS primary and stand-by instance deployment and increasing the number of EC2 instances.

However, this introduces additional complexity to our deployment, as we are about to spend the rest of the chapter covering the playbook, which will be deploying the resources. I want to keep this playbook as simple as possible for now.

## WordPress considerations and high availability

So far, we have been launching WordPress on a single server, which is fine. Still, as we are trying to remove as many of the single points of failure within our deployment as possible, we must put a little thought into how we initially configure and launch our deployment.

First, let's discuss the order we need to launch our deployment. The primary order in which we will need to tackle the elements is as follows:

- **VPC, subnets, internet gateway, routing, and security groups**: These are all needed to launch our deployment.

- **The Application Elastic Load Balancer**: We will be using the public hostname of the Elastic Load Balancer for our installation, so this needs to be launched before we start our installation.

- **The RDS database instance**: Our database instance must be available before we launch our installation, as we need to create the WordPress database and bootstrap the installation.

- **The EFS storage**: We need some storage to share between the EC2 instances we will be launching next.

So far, so good; however, this is where we have to start taking WordPress into account.

As some of you may know from experience, the current version of WordPress is not designed to be spread across multiple servers. We can apply plenty of hacks and workarounds to make WordPress play nicely in this sort of deployment; however, this chapter is about something other than the finer points of deploying WordPress. Instead, it is about using Ansible to deploy a multi-tiered web application.

Because of this, we will be going for the most basic of the multi-instance WordPress options by deploying our code and content on the EFS volume. This means that all we must do is install our LEMP stack. It should be noted that this option could be more performant at a large scale, but it will serve our needs.

Now, back to the list of tasks. When it comes to launching our instances, we need to do the following:

1. Launch a temporary EC2 instance running Ubuntu to reuse parts of existing playbooks.

2. Update the operating system and install the software stack, supporting tools, and configuration needed for us to install and run our WordPress installation.

3. Mount the EFS volume, set the correct permissions, and configure it to mount when the instance boots.

4. Bootstrap WordPress itself.

5. Create an AMI from our temporary instance and then terminate the temporary instance as it will not be needed now.

6. Create a launch template that uses the AMI we just created.

7. Create an autoscaling group and attach the launch configuration; it should also register our WordPress instances with the Elastic Load Balancer.

Further playbook runs, which will update the operating system and non-WordPress configuration, should repeat the process with the existing instances up and running, and then, once the AMI is built, it should be deployed alongside the current instances, which will then be terminated once the new instances are registered with the Elastic Load Balancer and receiving traffic.

This will allow us to update our operating system packages and configurations without downtime if everything goes as planned!

Now that we have an idea of what we are trying to achieve, let's make a start on our playbook.

# The Playbook

We will use the Playbook we looked at in *Chapter 10, Building Out a Cloud Network*, as a starting point, as all the roles are relevant to our deployment, and it already has the structure we need for our playbook.

We will also be using the roles to deploy and configure WordPress and the supporting software stack we used in *Chapter 9, Moving to the Cloud*, with a few tweaks, which are needed as we are targeting AWS and not Microsoft Azure; I will let you know when we get to them.

Unlike previous chapters, we will first look at the `site.yml` file to get an idea of the order in which we will run the roles.

There are three stages in the file, starting with the stage that deploys and configures our underlying AWS resources:

```yaml
- name: "Deploy and configure the AWS Environment"
  hosts: localhost
  connection: local
  gather_facts: true
  vars:
    state: "present"

  vars_files:
    - group_vars/common.yml

  roles:
    - vpc
    - subnets
    - gateway
    - securitygroups
    - elb
    - efs
    - rds
    - ec2tmp
    - endpoints
```

As you can see, this is the same as the `site.yml` file from *Chapter 10, Building Out a Cloud Network*, with additional roles added to the list from the `securitygroups` role downwards.

By the time our Playbook run gets to the second stage:

```yaml
- name: "Install and configure Wordpress"
  hosts: vmgroup
  gather_facts: true
  become: true
  become_method: "ansible.builtin.sudo"
```

```
    vars_files:
      - group_vars/common.yml
      - group_vars/generated_aws_endpoints.yml

    roles:
      - stack_install
      - stack_config
      - wordpress
```

A file called `group_vars/generated_aws_endpoints.yml` will have been generated, and there should be a temporary virtual machine instance up and running, meaning SSH should be accessible to the host running the Playbook.

Once this stage has been completed, our temporary virtual machine instance should have our software stack installed. WordPress will be freshly installed if this is the first time the playbook has been run, or if the playbook has detected an existing WordPress installation and left it alone unless there have been any changes to the plugin configuration from within the playbook.

The final stage is then run:

```
  - name: "Create AMI and update the Auto Scaling Group"
    hosts: localhost
    connection: local
    gather_facts: true
    vars:
      state: "present"

    vars_files:
      - group_vars/common.yml

    roles:
      - ec2ami
      - autoscaling
```

This stage creates an AMI from the temporary virtual machine instance, terminates the temporary instance as we no longer need it, creates a new version of our launch template, and then creates/updates the Auto Scaling Group to deploy the new version on the EC2 instances.

Sounds simple? Well, let's find out.

## The variables

Out of the box, there is a single variables file called `group_vars/common.yml` that contains all the static variables needed to deploy our environment.

Some additional files will be created in the `group_vars` folder throughout the Playbook run; they will contain some dynamically generated resources, such as passwords, resource names/endpoints, and other information.

We will discuss these files in more detail when we look at the tasks that create and interact with them; for now, we will look at the static variables defined within `group_vars/common.yml`, starting with the base application configuration.

### Application and resource configuration

We start the configuration with the option to enable/disable debug when running the Playbook. By default, it is set to `false`; however, when running the Playbook, I recommend switching it to `true` and reviewing the output:

```
debug_output: false
```

Next, we have the application name, region, and environment reference:

```
app:
  name: "learnansible"
  region: "eu-west-1"
  env: "prod"
```

The next block of variables defines details for the WordPress database; as we will be using the Amazon RDS service, we are just using the variables that are defined later in the file, so we only have to update the information in one place:

```
wp_database:
  name: "{{ rds.db_name }}"
  username: "{{ rds.db_username }}"
  password: "{{ rds.db_password }}"
```

The next block is the various variables used to configure WordPress itself:

```
wordpress:
  domain: "http://{{ aws_endpoints.elb }}/"
  title: "WordPress installed by Ansible on {{ os_family }}"
  username: "ansible"
  password: "{{ rds.db_password }}"
  email: "test@test.com"
  plugins:
    - "jetpack"
    - "wp-super-cache"
    - "wordpress-seo"
    - "wordfence"
    - "nginx-helper"
```

There are no significant changes to when we last defined these in *Chapter 9, Moving to the Cloud*, apart from using the `aws_endpoints.lb` variable, which won't be known until the Elastic Load Balancer has been launched. Also, for ease of use, we are reusing the password, which will be dynamically generated later in the file, as the WordPress admin password.

## Stack configuration

The next section overrides the defaults in the `roles/stack_install` role:

```
stack_packages:
  - "nginx"
  - "mariadb-client"
  - "php-cli"
  - "php-curl"
  - "php-fpm"
  - "php-gd"
  - "php-intl"
  - "php-mbstring"
  - "php-mysql"
  - "php-soap"
  - "php-xml"
  - "php-xmlrpc"
  - "php-zip"
  - "nfs-common" # Added for AWS
  - "nfs4-acl-tools" # Added for AWS
  - "autofs"  # Added for AWS
  - "rpcbind"  # Added for AWS
```

We have removed `mariadb-server` from the list of packages as we no longer need to install or configure a local database server, and we have added four packages at the end (all labeled `# Added for AWS`). These packages install the software required to mount the EFS filesystem using the NFS protocol, which leads us nicely into the next block:

```
nfs:
  mount_point: "/var/www/"
  mount_options:
 "nfsvers=4.1,rsize=1048576,wsize=1048576,hard,timeo=600,retrans=2"
  state: "mounted"
  fstype: "nfs4"
```

As you can see, this defines some basic information on where the EFS filesystem should be mounted, with what options and the type of filesystem it is.

### Resource names

This next section builds up the names of the resources we are going to be deploying; there is nothing too special happening here – it is just defined like this, so we don't have to update repeated information in several places manually:

```
vpc_name: "{{ app.name }}-{{ app.env }}-{{ playbook_dict.vpc }}"
internet_gateway_name: "{{ app.name }}-{{ app.env }}-{{ playbook_dict.
internet_gateway }}"
internet_gateway_route_name: "{{ internet_gateway_name }}-{{ playbook_
dict.route }}"
elb_target_group_name: "{{ app.name }}-{{ app.env }}-{{ playbook_dict.
elb_target_group }}"
elb_name: "{{ app.name }}-{{ app.env }}-{{ playbook_dict.elb }}"
efs_name: "{{ app.name }}-{{ app.env }}-{{ playbook_dict.efs }}"
rds_name: "{{ app.name }}-{{ app.env }}-{{ playbook_dict.rds }}"
ec2_tmp_name: "{{ app.name }}-tmp-{{ playbook_dict.ec2 }}"
ami_name: "{{ app.name }}-{{ app.env }}-{{ playbook_dict.ami }}"
ec2_name: "{{ app.name }}-{{ app.env }}-{{ playbook_dict.ec2 }}"
launch_template_name: "{{ app.name }}-{{ app.env }}-{{ playbook_dict.
lt }}"
asg_name: "{{ app.name }}-{{ app.env }}-{{ playbook_dict.asg }}"
```

We will not be covering the full `playbook_dict` block here as there is not much to see, although as a reminder, this is what the start of it looks like:

```
playbook_dict:
  deployedBy: "Ansible"
  ansible_warning: "Resource managed by Ansible"
  vpc: "vpc"
```

It just continues defining service names. The following section is where we start to define the variables used for the AWS resource deployment.

## EC2 configuration

The `ec2` variable is split into a few different layers. Layers for the auto-scaling group, the AMI, and the SSH keypair follow some general settings:

```
ec2:
  instance_type: "t2.micro"
  public_ip: true
  ssh_port: "22"
```

The variables are used across instances apart from the `public_ip` reference, which is only used when launching the temporary virtual machine instance to bootstrap WordPress.

The next layer defines some details about the auto-scaling group and launch template when used; they help define how many instances are launched, how updated instances are rolled out, and also, how the load balancer will check to see if they are healthy:

```
asg:
  min_size: 1
  max_size: 3
  desired_capacity: 2
  health_check_type: "EC2"
  replace_batch_size: 1
  health_check_period: 300
  replace_all_instances: true
  wait_for_instances: true
  wait_timeout: 900
  disable_api_termination: true
```

Next, we define the details about the base AMI we will use; as you can see, we are using Ubuntu 22.04, which is supplied by Canonical, the publisher and maintainer of Ubuntu:

```
ami:
  owners: "099720109477"
  filters:
    name: "ubuntu/images/hvm-ssd/ubuntu-jammy-22.04-amd64-server-*"
    virtualization_type: "hvm"
```

Finally, we have some details on the keypair to upload to AWS and use when launching our Virtual Machine instances:

```
keypair:
  name: "ssh_keypair"
  key_material: "{{ lookup('file', '~/.ssh/id_rsa.pub') }}"
```

Next up are the variables used when launching the RDS service.

### RDS configuration

These are all standard, apart from the rds.db_password variable:

```
rds:
  db_username: "{{ app.name }}"
  db_password: "{{ lookup('password', 'group_vars/generated_rds_
passwordfile chars=ascii_letters,digits length=30') }}"
  db_name: "{{ app.name }}"
```

```
    instance_type: "db.t2.micro"
    engine: "mysql"
    engine_version: "8.0"
    allocated_storage: "5"
```

As you can see, we are using a lookup module to add a random password to the `group_vars/generated_rds_passwordfile` file; we are instructing the module to generate a 30-character random password comprising letters and numbers only.

### EFS configuration

Here, we define the variables used to tell Ansible to wait and how long when creating the EFS resource:

```
efs:
  wait: "yes"
  wait_time: "1200"
```

### VPC and subnet configuration

This block remains unchanged from *Chapter 10, Building Out a Cloud Network*.

### Security group configuration

Most of this block is unchanged from *Chapter 10, Building Out a Cloud Network*, as we now define the SSH port as `ec2.ssh_port`. I have updated the EC2 group to use this reference rather than hardcoding port 22 into the block. The only other addition is the following:

```
elb_seach_string: "elb"
ec2_seach_string: "ec2"
rds_seach_string: "rds"
efs_seach_string: "efs"
```

These will be used throughout the playbook when we query the AWS API for information on our security groups.

### The final block

As per *Chapter 10, Building Out a Cloud Network*, this contains the following:

```
region: "{{ app.region }}"
```

That concludes our whistle-stop tour of the `group_vars/common.yml` file; as you can see, structure- and content-wise, we are following the same patterns as the last few chapters, where we group variables into logical blocks and trying to reuse references as much as possible throughout so that we don't have to repeat information repeatedly.

## The Playbook roles

Now that we have covered the variables, we can work through the roles in the order they appear in the site.yml file.

### The VPC, subnets, gateway, and security groups roles

There are no changes to these roles from *Chapter 10, Building Out a Cloud Network*; they are just dropped in place and work as expected. The remaining roles in this section of the Playbook will reference the output of these roles when referring to subnets, security groups, and the VPC.

### The Application Elastic Load Balancer (ELB) role

In this role, we will deploy two resources, the first of which is a target group. This will be used when we launch our auto-scaling virtual machine instances – we attach our instances to the target group. Then, the target group is attached to the Application Elastic Load Balancer, which we will also launch in this role.

The task itself is pretty static, as you can see from the code for the following task:

```
- name: "Provision the target group"
  community.aws.elb_target_group:
    name: "{{ elb_target_group_name }}"
    region: "{{ region }}"
    state: "{{ state }}"
    protocol: "http"
    port: "80"
    deregistration_delay_timeout: "15"
    vpc_id: "{{ vpc_output.vpc.id }}"
    modify_targets: "false"
    tags:
      "Name": "{{ elb_target_group_name }}"
      "projectName": "{{ app.name }}"
      "environment": "{{ app.env }}"
      "deployedBy": "{{ playbook_dict.deployedBy }}"
      "description": "{{ playbook_dict.ansible_warning }}"
      "role": "target-group"
  register: elb_target_group_output
```

We are just referencing variables, with the only dynamic content being the ID of the VPC, which is referenced from the vpc_output variable we registered when launching the VPC in the VPC role.

As we are registering some output in this role, we will continue by adding a debug task straight after; in this case, the task looks like the following:

```
- name: "Debug: ELB Target Group Output"
  ansible.builtin.debug:
    var: "elb_target_group_output"
  when: debug_output
```

As we have already covered in *Chapter 10, Building Out a Cloud Network*, we will not be repeating these tasks in our overview of the Playbook unless we are doing something different – so, from now on, if we are registering an output, please assume that a debug task immediately follows.

There is one more bit of information we need before we create the ELB, and that's the ID of the security group.

To get this, we can loop through the `security_groups_with_rules_output` variable and use `set_fact` to set the `group_id` when the `group_name` contains the contents of the `elb_seach_string` variable:

```
- name: Extract ELB Group ID
  ansible.builtin.set_fact:
    elb_group_id: "{{ item.group_id }}"
  loop: "{{ security_groups_with_rules_output.results }}"
  when: item.group_name is search(elb_seach_string)
```

Whenever we need the ID of a security group, we will use this same pattern but update the name of the fact that is being set and the corresponding search steering variable.

The following task provisions the Application Elastic Load Balancer, which will be used to distribute HTTP requests across our auto-scaling managed virtual machine instances to serve our WordPress site:

```
- name: "Provision an application elastic load balancer"
  amazon.aws.elb_application_lb:
    region: "{{ region }}"
    name: "{{ elb_name }}"
    state: "{{ state }}"
    security_groups: "{{ elb_group_id }}"
    subnets: "{{ subnet_public_ids }}"
    listeners:
      - Protocol: "HTTP"
        Port: "80"
        DefaultActions:
          - Type: "forward"
            TargetGroupArn: "{{ elb_target_group_output.target_group_
arn }}"
    tags:
```

```
        "Name": "{{ elb_name }}"
        "projectName": "{{ app.name }}"
        "environment": "{{ app.env }}"
        "deployedBy": "{{ playbook_dict.deployedBy }}"
        "description": "{{ playbook_dict.ansible_warning }}"
        "role": "load-balancer"
  register: loadbalancer_output
```

As you can see, we are attaching the Application Elastic Load Balancer to the subnets defined listed in the subnet_public_ids, and we are attaching the security group with the ID defined in the elb_group_id fact that registered in the previous task.

We are then configuring a listener on port 80 to accept HTTP traffic and forward it to the Target Group we launched at the start of the role – which concludes the Application Elastic Load balancer role.

### The Elastic File System (EFS) role

The role starts with the task which sets the efs_group_id using the efs_seach_string variable. Once we know the ID of the security group we are applying to the EFS service, we can move on to the next task.

This task generates a file using a template and places it in the group_vars folder:

```
- name: "Generate the efs targets vars file"
  ansible.builtin.template:
    src: "targets.j2"
    dest: "group_vars/generated_efs_targets.yml"
    mode: "0644"
```

The template file used to populate the file at group_vars/generated_efs_targets.yml looks like the following:

```
efs_targets:
{% for item in subnet_storage_ids %}
      - subnet_id: "{{ item }}"
        security_groups: [ "{{ efs_group_id }}" ]
{% endfor %}
```

Here, we are using a Jinja2 for loop to loop through the contents of subnet_storage_ids, which will create a file that looks something like the following:

```
efs_targets:
      - subnet_id: "subnet01_id"
        security_groups: [ "efs_group_id" ]
      - subnet_id: "subnet02_id"
        security_groups: [ "efs_group_id" ]
```

```
    - subnet_id: "subnet03_id"
      security_groups: [ "efs_group_id" ]
```

This means that when we create the EFS file system, it will be available across all the availability zones in our chosen region.

Well, it will be once we load in the contents of the file we have just loaded, which we do in the next task, as you can see here:

```
- name: "Include the efs targets vars file"
  ansible.builtin.include_vars: "group_vars/generated_efs_targets.yml"
```

We now have everything in place to create the EFS file system, which is done using this task:

```
- name: "Create the EFS File System"
  community.aws.efs:
    name: "{{ efs_name }}"
    region: "{{ region }}"
    state: "{{ state }}"
    tags:
      "Name": "{{ efs_name }}"
      "projectName": "{{ app.name }}"
      "environment": "{{ app.env }}"
      "deployedBy": "{{ playbook_dict.deployedBy }}"
      "description": "{{ playbook_dict.ansible_warning }}"
      "role": "efs"
    targets: "{{ efs_targets }}"
    wait: "{{ efs.wait }}"
    wait_timeout: "{{ efs.wait_time }}"
  register: efs_output
```

It can take a few minutes to create the file system, and we must wait until this task has succeeded before we continue, which is why we are using the wait flag. If we don't wait, we increase the risk that the file system will not be ready by the time our virtual machine is launched and unable to mount it, which will cause the Playbook execution to fail.

Speaking of tasks that take a while, the next role deals with launching the Amazon RDS instance, which we will use as the database for our WordPress site. This task can take up to 10 minutes to complete.

### The Amazon RDS role

There are two main parts to the role; the first does a similar task to the one we had to do in the previous role when we created the targets for the EFS to be attached to.

The RDS service differs in that rather than passing in the subnets manually when we deploy the service, we can create a group natively on the AWS side and then reference it when we launch the RDS instance.

The task to create the RDS subnet group looks like the following:

```
- name: "Add RDS subnet group"
  amazon.aws.rds_subnet_group:
    name: "{{ rds_name }}"
    region: "{{ region }}"
    state: "{{ state }}"
    description: "{{ dict.ansible_warning }}"
    subnets: "{{ subnet_database_ids }}"
    tags:
      "Name": "{{ rds_name }}"
      "projectName": "{{ app.name }}"
      "environment": "{{ app.env }}"
      "deployedBy": "{{ playbook_dict.deployedBy }}"
      "description": "{{ playbook_dict.ansible_warning }}"
      "role": "rds"
  register: rds_subnet_group_output
```

Once we have created the subnet group, we need to find the security group ID using the `rds_seach_string` variable and set a fact called `rds_group_id`.

Now we have all the information we need to launch the RDS instance, the task for which looks like the following:

```
- name: "Create the RDS instance"
  amazon.aws.rds_instance:
    id: "{{ rds_name }}"
    region: "{{ region }}"
    state: "{{ state }}"
    db_instance_class: "{{ rds.instance_type }}"
    engine: "{{ rds.engine }}"
    engine_version: "{{ rds.engine_version }}"
    allocated_storage: "{{ rds.allocated_storage }}"
    username: "{{ rds.db_username }}"
    password: "{{ rds.db_password }}"
    db_name: "{{ rds.db_name }}"
    db_subnet_group_name: "{{ rds_subnet_group_output.subnet_group.
name }}"
    vpc_security_group_ids: ["{{ rds_group_id }}"]
    tags:
      "Name": "{{ rds_name }}"
      "projectName": "{{ app.name }}"
      "environment": "{{ app.env }}"
      "deployedBy": "{{ playbook_dict.deployedBy }}"
      "description": "{{ playbook_dict.ansible_warning }}"
```

```
        "role": "rds"
    register: rds_instance_output
```

As mentioned at the end of the previous task, this can take quite a while to deploy, typically just over 10 minutes, so when we run the Playbook, this task will appear to have stalled.

So please do not worry – it is busy working away in the background.

Once this role has finished running, we will have all the core AWS resources we need to launch an EC2 instance, perform the software configuration, and install WordPress.

### The temporary EC2 instance role

Before we work through the tasks that launch the temporary instance, let's go into a little more detail on why we need a temporary EC2 instance in the first place.

As we mentioned in the introduction, this instance will be running Ubuntu, and we will be targeting it with slightly modified copies of the `stack_install`, `stack_config`, and `wordpress` roles that we first ran locally in *Chapter 5, Deploying WordPress*, and against a single cloud instance in *Chapter 9, Moving to the Cloud*.

One of the modifications we will be making to the roles is installing the software needed to mount our EFS, which we will then use to store the WordPress code and supporting files for our WordPress installation, meaning that we have everything we need file-wise for WordPress on a shared file system we can then mount on multiple virtual machine instances.

The second change is that rather than installing a database server on our local instance, we will be using the Amazon RDS database service for WordPress, meaning that we can have multiple instances of WordPress, all being able to connect to a single remote database.

Great, you may be thinking to yourself, but that doesn't explain why this is a temporary instance.

Well, once everything has been installed, mounted, configured, and WordPress bootstrapped, we will be making our own **Amazon Machine Image** (**AMI**) and terminating the temporary EC2 instance. Once it's been terminated, we will take the AMI and configure our Auto Scaling Group to use the newly created image, which will either trigger the deployment of new hosts if it is our first time running the Playbook or it will launch more instances and terminate the old ones if we have already had virtual machine instances running our WordPress installation.

When these virtual machine instances boot up using our custom AMI, they will already have NGINX and PHP installed and configured, ready to serve WordPress, and the EFS containing our WordPress files will be mounted, meaning that our servers will be good to go as soon as they are deployed.

All of this means our WordPress installation should be sound to scale up if we have an influx of traffic hitting the site for whatever reason, and all instances of our virtual machines will be running a known good configuration; in fact, it will be the same configuration as the other hosts serving our WordPress site.

Just as important, as we are not relying on anything on the local virtual machine instances filesystem, we are just as good at automatically scaling down by terminating hosts automatically when the influx of traffic has subsided without the risk of data loss or availability.

If this approach is planned right – in theory, we don't even need SSH access to the hosts launched by the Auto Scaling Group as we should never need to manage them manually, and we can treat them as short-lived instances where we don't have to care if they are running or terminated – just that we have the desired of instances delivering our application.

So, now that we know why we are taking this approach, let's return to the Playbook and look at the tasks needed to get this temporary EC2 instance up and running to the point where we can SSH to it and install our software and WordPress.

The first task is to get a list of all the Ubuntu AMIs using the variables we covered earlier in the chapter:

```
- name: "Gather information about AMIs with the specified filters"
  amazon.aws.ec2_ami_info:
    region: "{{ region }}"
    owners: "{{ ec2.ami.owners }}"
    filters:
      name: "{{ ec2.ami.filters.name }}"
      virtualization-type: "{{ ec2.ami.filters.virtualization_type }}"
  register: ubuntu_ami_info
```

The list of AMIs returned will contain all of the various AMI versions for our chosen Ubuntu version; we only need to know the ID of the latest version published by Canonical (the publisher and maintainer of Ubuntu) so we know we are using the most up-to-date image that contains the latest patches and any bug fixes.

Luckily, each AMI returned in the list has a key called `creation_date`, the value of which, as you may have guessed, is the date and time the AMI was published. This means we can run the following task to get the ID of the latest version of the AMI:

```
- name: "Filter the list of AMIs to find the latest one"
  ansible.builtin.set_fact:
    ami: "{{ ubuntu_ami_info.images | sort(attribute='creation_date')
| last }}"
```

As you can see, this takes the content of the list, which is defined as `ubuntu_ami_info.images`, sorts the list by `creation_date`, and then takes the ID of the `last` AMI in the list as, by default, they are sorted in ascending order.

Now that we know the ID of the most up-to-date Ubuntu AMI, we can progress with more preparation work before launching our EC2 instance.

We now need to create an SSH key pair on the AWS side. This will contain the public portion of the SSH key we will use to access the EC2 instance when it is launched – the task to configure this looks like the following and uses the variables we covered earlier in the chapter to get the contents of the public portion of our SSH key:

```
- name: "Create a SSH Key Pair"
  amazon.aws.ec2_key:
    region: "{{ region }}"
    state: "{{ state }}"
    name: "{{ ec2.keypair.name }}"
    key_material: "{{ ec2.keypair.key_material }}"
    tags:
      "Name": "{{ ec2.keypair.name }}"
      "projectName": "{{ app.name }}"
      "environment": "{{ app.env }}"
      "deployedBy": "{{ playbook_dict.deployedBy }}"
      "description": "{{ playbook_dict.ansible_warning }}"
      "role": "ssh_keypair"
  register: keypair_output
```

Finally, before we launch our EC2 instance, we need the ID of the security group, which allows the public IP address of our host running Ansible SSH access to the EC2 instance. To do this, we set a fact called `ec2_group_id` using the `ec2_seach_string` variable to find the correct group ID.

Now, we have everything in place to launch the EC2 instance using the following task:

```
- name: "Create the temporary ec2 instance"
  amazon.aws.ec2_instance:
    name: "{{ ec2_tmp_name }}"
    region: "{{ region }}"
    state: "{{ state }}"
    vpc_subnet_id: "{{ subnet_compute_ids[0] }}"
    instance_type: "{{ ec2.instance_type }}"
    security_group: "{{ ec2_group_id }}"
    key_name: "{{ ec2.keypair.name }}"
    network:
      assign_public_ip: "{{ ec2.public_ip }}"
    image_id: "{{ ami.image_id }}"
    tags:
      Name: "{{ ec2_tmp_name }}"
      Description: "{{ dict.ansible_warning }}"
      Project: "{{ app.name }}"
      Environment: "{{ app.env }}"
      Deployed_by: "Ansible"
```

```
        Role: "tmp"
    register: ec2_tmp_instance_output
```

The only thing pointed out in the preceding task is that when we add the value for the vpc_subnet_id we can only pass in a single ID. As we don't need this virtual machine instance to be highly available, that is not a problem, so we are using the first ID in the list of subnet IDs by using the {{ subnet_compute_ids[0] }}.

When launching an EC2 instance in AWS, it goes through a few stages and, by default, the amazon.aws.ec2_instance module creates the instance and doesn't wait for the status to change from *creating* to *running*.

Our next task polls the AWS API waiting for the status of our EC2 instance to be *running*:

```
- name: "Get information about the temporary EC2 instance to see if it
is running"
  amazon.aws.ec2_instance_info:
    region: "{{ region }}"
    filters:
      instance-id: "{{ ec2_tmp_instance_output.instances[0].instance_
id }}"
  register: ec2_tmp_instance_state
  delay: 5
  retries: 50
  until: ec2_tmp_instance_state.instances[0].state.name == "running"
```

As you can see, the previous task takes the ID of our newly created EC2 instance and polls the AWS API every 5 seconds, a maximum of 50 times, until the value of ec2_tmp_instance_state.instances[0].state.name is equal to running.

You might think to yourself that it seems a bit overkill to do that, and 99% of the time, you would be correct – it usually takes no more than a few checks for the status to change. Still, there is the odd occasion that AWS might be on a "go-slow," and during testing, I have seen it take up to 15 checks, or just over a minute, for the status to change, so we need to take this delay into account in our Playbook as it could break the Playbook execution if we don't.

The next task takes the details, the DNS name and IP address, of our now-running EC2 instance and adds them to the host group called vmgroup:

```
- name: "Add the temporary EC2 instance to the vmgroup"
  ansible.builtin.add_host:
    name: "{{ ec2_tmp_instance_output.instances[0].public_dns_name }}"
    ansible_ssh_host: "{{ ec2_tmp_instance_output.instances[0].public_
ip_address }}"
    groups: "vmgroup"
```

Before we hand off to the next role, we should perform one more check.

Sometimes, the Ansible Playbook works through the tasks so quickly that it is possible that even though our EC2 instance has a status of *running*, it does not mean that the host has finished booting, and SSH is started and is accessible:

```
- name: "Wait for the temporary EC2 instance to be ready to accept SSH
  connections"
  ansible.builtin.wait_for:
    host: "{{ ec2_tmp_instance_output.instances[0].public_ip_address
}}"
    port: "{{ ec2.ssh_port }}"
    delay: 10
    timeout: 300
```

Now that we have confirmation that our EC2 host is accessible to our machine running Ansible using SSH, we can proceed to the final role in this section of the site.yml file.

### The endpoints role

This role has a single task, which creates a file at generated_aws_endpoints.yml containing the name of the AWS endpoints for the EFS, RDS, and ELB resources we have created:

```
- name: "Generate the aws endpoints file"
  ansible.builtin.template:
    src: "endponts.j2"
    dest: "group_vars/generated_aws_endpoints.yml"
    mode: "0644"
```

The endponts.j2 template file looks like the following:

```
aws_endpoints:
  efs: "{{ efs_output.efs.filesystem_address.split(':')[0] }}"
  rds: "{{ rds_instance_output.endpoint.address }}"
  elb: "{{ loadbalancer_output.dns_name }}"
```

Both the RDS and ELB endpoints are straightforward enough; for the EFS, you might notice something at the end – what is that for?

None of the output that is registered under the efs_output.efs variable contains just the address of the EFS endpoint. The one we are using, filesystem_address, has information on the file system mount, which is represented by appending :/ to the end of the DNS address we need.

To get around this, we are using the split function, passing : as the delimiter and then taking the first section, which is defined as 0, meaning that we end up with everything before the :, which is the DNS name we are after.

Now that we have a populated `group_vars/generated_aws_endpoints.yml` file, we can load it into the second section of the `site.yml` file as a variable file, saving us from having to interact with the AWS from our EC2 instance.

So, now that we have our EC2 instance up and running, let's get our software stack installed, configured, and WordPress bootstrapped.

### The stack install role

The tasks in this role remain unchanged from the previous times we have executed the Playbook because all the changes we have made are in the `stack_packages` variable we are passing in.

As a reminder, this role does the following:

- Updates the APT cache and ensures that the installed packages are running the latest available versions – which shouldn't be too many as we are using the newest AMI

- Imports the APT keys for the additional repositories we will be enabling

- Installs the packages containing details of the additional repositories and enables them

- Installs the packages listed in the `system_packages`, `extra_packages`, and `stack_packages` variables – `system_packages` and `extra_packages` contain the default values we have been using throughout, and because we are passing the updated `stack_packages` variable via the `group_vars/common.yml` file, this overrides the default values from previous chapters which are still defined in the `roles/stack_install/defaults/main.yml` file

This leaves us with all the base software we need to install on the EC2 instance.

### The stack configuration role

Unlike the previous role, there are some amendments to this role, starting with additional tasks out of the gate.

Three tasks are added to the top of `roles/stack_config/tasks/main.yml`, the first of which is a continuation of the checks we did towards the end of the roles in the last section of the `site.yml` file:

```
- name: "Check that the EFS volume is ready"
  ansible.builtin.wait_for:
    host: "{{ aws_endpoints.efs }}"
    port: "2049"
    delay: 10
    timeout: 300
```

As you can see, this checks that port 2049 is accessible at the endpoint defined in aws_endpoints. efs; the reason why this is there is that while the EFS service is ready, it may take a little while for the DNS records for the endpoint to be updated and accessible within the VPC. As we will soon attempt to mount the EFS filesystem, we must ensure it is accessible before proceeding.

The next task is to ensure that the RPC Bind service is up and running; we will need to mount the EFS file system:

```
- name: "ensure rpcbind service is running"
  ansible.builtin.service:
    name: "rpcbind"
    state: "started"
    enabled: true
```

The final additional task mounts the EFS and ensures that it is added to the file system configuration to ensure that from now on, the EFS is mounted when the EC2 instance boots:

```
- name: "mount the EFS volume"
  ansible.posix.mount:
    src: "{{ aws_endpoints.efs }}:/"
    path: "{{ nfs.mount_point }}"
    opts: "{{ nfs.mount_options }}"
    state: "{{ nfs.state }}"
    fstype: "{{ nfs.fstype }}"
```

As you will have already seen from when we covered the variables at the start of the chapter, we are mounting the EFS at /var/www/; we are making sure to do this before the following two tasks to ensure that our WordPress users home directory is created on the share.

These two tasks remain unchanged from the last time we installed WordPress, as does the value of wordpress_system.home, which is /var/www/wordpress.

So, now that we have created our WordPress user and group, we can proceed with the rest of the tasks:

- Update /etc/nginx/nginx.conf with some sensible defaults

- Create the configuration for our default host at /etc/nginx/conf.d/default.conf

- Create the /etc/nginx/global directory and copy the restrictions.conf and wordpress_shared.conf files there

The next task is more of a quality-of-life improvement to do with the way our Playbook deals with PHP, as this Playbook is designed to keep our WordPress installation up to date by taking the base Ubuntu image and bootstrapping from scratch each time rather than managing the configuration in place. It is possible that the version of PHP could change at some point during the life of our WordPress installation.

So far, whenever the `stack_config` role has been executed, it has been using the following variables:

```
php_fpm_path: "/etc/php/8.1/fpm/pool.d/www.conf"
php_ini_path: "/etc/php/8.1/fpm/php.ini"
php_service_name: "php8.1-fpm"
```

As you can see, 8.1 is a hardcoded value. While we can overwrite these variables at the variable level elsewhere in our configuration, it would be better to work out which version of PHP is installed at runtime and reference that.

To do this, we can update these values as follows:

```
php_fpm_path: "/etc/php/{{ php_version }}/fpm/pool.d/www.conf"
php_ini_path: "/etc/php/{{ php_version }}/fpm/php.ini"
php_service_name: "php{{ php_version }}-fpm"
```

This means we now must find a way to populate the `php_version` variable with the relevant version of PHP.

To do this, we can run the `php -v` command, which returns a lot of information on the version of PHP installed. We then use the `head` and a few `cut` commands on the Linux command line using the `ansible.builtin.shell` and not a built-in Ansible function:

```
- name: "Get the PHP version"
  ansible.builtin.shell:
    cmd: "php -v | head -n 1 | cut -d ' ' -f 2 | cut -c 1-3"
  register: php_version_output
```

Here is a detailed breakdown of the command we are getting Ansible to run:

- `php -v`: This command, when run, outputs the version information of the PHP installed on the host the command is being executed on; this output is typically a multi-line text that includes the PHP version along with additional information on how the version of the PHP was compiled.

- `|`: This symbol is known as a pipe. It takes the command output on its left (in this case, `php -v`) and uses it as the input for the command on its right. It's a way of passing data between programs.

- `head -n 1`: This command processes the input received from the previous command; the head command outputs the first part of the files or data it receives. `-n 1` is an option that tells head to output only the first line. So, in our case, `head -n 1` takes the multiple lines of output from `php -v` and returns just the very first line.

- `|`: Another pipe, which again passes the command output on its left, `head -n 1`, to the command on its right.

- `cut -d ' ' -f 2`: This command is used for cutting out sections of each input line. `-d ' '` is an option where `-d` stands for the delimiter, and `' '` (a space) is the delimiter being used. This tells cut to divide each line into sections based on spaces. `-f 2` means *field 2*. This option tells the `cut` command to select the second field of the line in the standard format of the PHP version output; this field should be the version number.

- `|`: Again, we have another pipe, passing the output, now just the version number, to the following command.

- `cut -c 1-3`: This further processes the version number. `-c 1-3` tells cut to return only the characters in positions 1 through 3 of the string it receives. For a typical PHP version such as 8.2.1, this would result in 8.2, which is precisely what we need to proceed with the rest of our tasks.

We can then take the output and register it as `php_version_output`, and set the `php_version` variable as a fact:

```
- name: "Set the PHP version"
  ansible.builtin.set_fact:
    php_version: "{{ php_version_output.stdout }}"
```

Now that we have the PHP version, we can proceed with the remainder of the PHP tasks, which copy the `www.conf` file to `/etc/php/{{ php_version }}/fpm/pool.d/www.conf` and also update the `PHP.ini` file at `/etc/php/{{ php_version }}/fpm/php.ini`.

With those files in place, we start the PHP-FPM and NGINX services, ensuring that they are set to start on boot.

The final task in the role is to create the `~/.my.cnf` file and populate it with the information of our Amazon RDS instance. All of the other MariaDB tasks, which are there to start and configure our local MariaDB server, are commented out as we no longer install a local database server, so we don't need to run the tasks to configure it.

### The WordPress role

There are just two tasks commented out in this role. The tasks that create the database and the database user are not needed because when the Amazon RDS instance started, the database and user were made for us, meaning these two tasks are redundant.

All other tasks remain; for more details, see *Chapter 5, Deploying WordPress*.

### The EC2 AMI role

Now that our software stack is installed and configured and WordPress is sorted, it is time to create the AMI from our temporary instance.

The first thing we need to do is get the details on our temporary EC2 instance; as our host group contains the DNS name of the instance, we can use this:

```
- name: "Find out some facts about the instance we have been using"
  amazon.aws.ec2_instance_info:
    region: "{{ region }}"
    filters:
      dns-name: "{{ groups['vmgroup'] }}"
  register: our_instance
```

Now that we have the information on the instance we would like to create the AMI from registered as our_instance, we can proceed with the AMI creation:

```
- name: "Create the AMI"
  amazon.aws.ec2_ami:
    region: "{{ region }}"
    state: "{{ state }}"
    instance_id: "{{ our_instance.instances[0].instance_id }}"
    wait: "yes"
    name: "{{ ami_name }}-{{ ansible_date_time.date }}_{{ ansible_
date_time.hour }}{{ ansible_date_time.minute }}"
    tags:
      "Name": "{{ ami_name }}-{{ ansible_date_time.date }}_{{ ansible_
date_time.hour }}{{ ansible_date_time.minute }}"
      "buildDate": "{{ ansible_date_time.date }} {{ ansible_date_time.
time }}"
      "projectName": "{{ app.name }}"
      "environment": "{{ app.env }}"
      "deployedBy": "{{ playbook_dict.deployedBy }}"
      "description": "{{ playbook_dict.ansible_warning }}"
      "role": "{{ playbook_dict.ami }}"
  register: ami_output
```

There are just a few things to point out here. As you can see, we are using ansible_date_time to generate the date and get the current time as an hour and minute. We are using this both to give a unique name for the AMI and add a tag called buildDate.

The reason why we are using both the date and time is that it could be possible that we will need to create multiple AMIs on a single day, so it is important that we can easily identify them by name.

Once the AMI is created, we do not need the temporary instance, so we can terminate it:

```
- name: "Remove any temporary  instances which are running"
  amazon.aws.ec2_instance:
    region: "{{ region }}"
    state: "absent"
```

```
    name: "{{ ec2_tmp_name }}"
    filters:
      instance-state-name: "running"
      "tag:Name": "{{ ec2_tmp_name }}"
      "tag:Role": "tmp"
      "tag:Project": "{{ app.name }}"
```

Once the EC2 instance has been terminated, there is one more task in the role:

```
- name: "Wait for 2 minutes before continuing"
  ansible.builtin.pause:
    minutes: 2
```

This does exactly what it says: it pauses the Playbook execution for 2 minutes.

I have included this because there was the odd occasion where the AMI was created and shown as available. Still, for some reason, it takes a short while for it to appear in the results when we query the Amazon API to find our AMIs, so rather than introduce a potential error when the next role starts, I have found it best to wait a minute or two.

### The auto-scaling role

We have arrived at the final role of the Playbook; in this role, we will create all the resources needed to deploy EC2 instances using our newly created AMI and register them with the ELB to access our WordPress site.

The first thing we need to do is grab a list of all our AMIs from the API:

```
- name: "Search for all of our AMIs"
  amazon.aws.ec2_ami_info:
    region: "{{ region }}"
    filters:
      name: "{{ ami_name }}-*"
  register: ami_find
```

Now that we have a list of AMIs, we need to filter out the most recent one. To do this, we use the same logic that we used when launching the temporary EC2 instance:

```
- name: "Find the last one we built"
  ansible.builtin.set_fact:
    ami_sort_filter: "{{ ami_find.images | sort(attribute='creation_
date') | last }}"
```

Now that we have filtered our list of AMIs down to the latest one, we need to set two facts, one for the name of the AMI and the other containing the ID of the AMI:

```
- name: "Grab AMI ID and name of the most recent result"
  ansible.builtin.set_fact:
    our_ami_id: "{{ ami_sort_filter.image_id }}"
    our_ami_name: "{{ ami_sort_filter.name }}"
```

The final bit of information we need before we start creating/updating resources is the ID of the security group we are using for the EC2 instances.

As before, we use the `ec2_seach_string` variable to find the correct group ID and set a fact called `ec2_group_id`.

Next up, we need to create or update a launch template if one already exists.

A launch template contains the basic configuration for the instances we will be launching in the auto-scaling group:

```
- name: "Create the launch template"
  community.aws.ec2_launch_template:
    region: "{{ region }}"
    state: "{{ state }}"
    name: "{{ launch_template_name }}"
    version_description: "{{ our_ami_name }}"
    image_id: "{{ our_ami_id }}"
    security_group_ids: ["{{ ec2_group_id.security_groups[0].group_id
}}"]
    instance_type: "{{ ec2.instance_type }}"
    disable_api_termination: "{{ ec2.asg.disable_api_termination }}"
    tags:
      "Name": "{{ ec2_name }}"
      "projectName": "{{ app.name }}"
      "environment": "{{ app.env }}"
      "deployedBy": "{{ playbook_dict.deployedBy }}"
      "description": "{{ playbook_dict.ansible_warning }}"
      "role": "launchTemplate"
```

With this task, we create the launch template and then publish a version called after the name of our AMI so that we can quickly identify it; we then attach the corresponding AMI ID and security group ID and set the spec of the instances we want to launch.

With the launch template in place, we need to gather a few more bits of information from the AWS API before creating the auto-scaling group.

First, we need the ID of the target group that we created in the ELB role:

```
- name: "Find out the target group ARN"
  community.aws.elb_target_group_info:
    region: "{{ region }}"
    names:
      - "{{ elb_target_group_name }}"
  register: elb_target_group_output
```

We then need the IDs of the subnets where we are going to be deploying the EC2 instances launched as part of auto-scaling group, the following task gathers information on the subnets:

```
- name: "Get information on the ec2 subnets"
  amazon.aws.ec2_vpc_subnet_info:
    region: "{{ region }}"
    filters:
      tag:role: "*{{ subnet_role_compute }}*"
  register: ec2_subnet_output
```

Now that we have the information on the subnets, we need to extract just the IDs of each of the subnets and create a list:

```
- name: "Create a list of subnet IDs"
  ansible.builtin.set_fact:
    subnet_ec2_ids: "{{ subnet_ec2_ids | default([]) + [item.subnet_
id] }}"
  loop: "{{ ec2_subnet_output.subnets }}"
```

This is the final bit of information we need, and we can now proceed with creating or updating the auto-scaling group:

```
- name: "Create/update the auto-scaling group using the launch
template we just created"
  amazon.aws.autoscaling_group:
    region: "{{ region }}"
    state: "{{ state }}"
    name: "{{ asg_name }}"
    target_group_arns: ["{{ elb_target_group_output.target_groups[0].
target_group_arn }}"]
    launch_template:
      launch_template_name: "{{ launch_template_name }}"
    min_size: "{{ ec2.asg.min_size }}"
    max_size: "{{ ec2.asg.max_size }}"
    desired_capacity: "{{ ec2.asg.desired_capacity }}"
    health_check_period: "{{ ec2.asg.health_check_period }}"
    health_check_type: "{{ ec2.asg.health_check_type }}"
```

```
    replace_all_instances: "{{ ec2.asg.replace_all_instances }}"
    replace_batch_size: "{{ ec2.asg.replace_batch_size }}"
    vpc_zone_identifier: "{{ subnet_ec2_ids }}"
    wait_for_instances: "{{ ec2.asg.wait_for_instances }}"
    wait_timeout: "{{ ec2.asg.wait_timeout }}"
    tags:
      - key: "Name"
        value: "{{ ec2_name }}"
        propagate_at_launch: true
      - key: "Project"
        value: "{{ app.name }}"
        propagate_at_launch: true
      - key: "Environment"
        value: "{{ app.env }}"
        propagate_at_launch: true
      - key: "Deployed_by"
        value: "Ansible"
        propagate_at_launch: true
  register: ec2_asg_output
```

There is quite a lot happening in this, the final resource we will be launching, so let's go into more detail.

First, we have the basic configuration standard across most of the AWS-related modules we have called throughout this Playbook; here, we are setting the name, region, and state of the resource, which will be present for this playbook.

Next up, we must provide the Target Group **Amazon Resource Names** (**ARNs**). The target_ group_arns key specifies the ARNs of the target groups for the load balancer, which we set to the first target group ARN from elb_target_group_output and then the launch_template key references the launch template by its name, set to the value of launch_template_name.

Now we have the size and capacity settings; the min_size, max_size, and desired_capacity keys are set using ec2.asg.min_size, ec2.asg.max_size, and ec2.asg.desired_ capacity variables, which define the auto-scaling group's minimum, maximum, and desired number of instances.

We then have the health check configuration, setting the health_check_period and health_ check_type keys to control how the health of the instances in the **auto scaling group** (**ASG**) is checked.

Now we have the Instance Replacement Settings. The replace_all_instances and replace_ batch_size keys instruct whether all instances should be replaced and provide the batch size for replacing instances, respectively.

Then, we have the Network Configuration, setting `vpc_zone_identifier` to use the list of subnet IDs stored in `subnet_ec2_ids` to distribute the instances in the ASG across those subnets and availability zones.

Next up are the Wait Settings, which control whether the task should wait for the instances to have a status of `running` and the maximum time to wait for that condition to be met.

Finally, you will have noticed that we are tagging in a pretty different way than we have been doing throughout the rest of the Playbook; the task defines several tags (`Name`, `Project`, `Environment`, and `Deployed_by`) with respective values, all marked to propagate at launch, which means that the EC2 instances launched by the auto-scaling group will each inherit these tags when they are launched.

This concludes our walk-through of the Playbook. As you will have seen, we extended our original AWS networking Playbook from *Chapter 10*, *Building Out a Cloud Network*, to encompass more services as well as integrating our WordPress roles from the Playbook we covered in *Chapter 5*, *Deploying WordPress* – all that is left now is run the playbook.

# Running the Playbook

Now that we have all the roles needed to deploy our resources into AWS, we can run the playbook. To start with, we need to let Ansible know our access key and secret by running the following commands with your own credentials to set the environment variables:

```
$ export AWS_ACCESS_KEY=AKIAI5KECPOTNTTVM3EDA
$ export AWS_SECRET_KEY=Y4B7FFiSWl0Am3VIFc07lgnc/TAtK5+RpxzIGTr
```

With environment variables set, you kick off the Ansible run by using the following command:

```
$ ansible-playbook -i hosts site.yml
```

Unlike previous chapters, where we just looked at the end of the playbook run, here we will look at some highlights of what happens when we deploy our resources.

## Playbook run highlights

This is not the complete playbook output, and when running the playbook, I have not enabled debug, so all those tasks will be skipped.

We start with the VPC:

```
PLAY [Deploy and configure the AWS Environment] **********
TASK [Gathering Facts] ********************************
ok: [localhost]
TASK [roles/vpc : Create VPC] ****************************
changed: [localhost]
```

We now have somewhere to put the subnets once we have gathered some information on the availability zones in our chosen region:

```
TASK [roles/subnets : Get some information on the available zones]
**************
ok: [localhost]
```

Once we have that information, it will loop through and include the create_subnet.yml tasks:

```
TASK [roles/subnets : Create all subnets] ******************
included: create_subnet.yml for localhost => (item={'name': 'ec2',
'role': 'compute'})
included: create_subnet.yml for localhost => (item={'name': 'rds',
'role': 'database'})
included: create_subnet.yml for localhost => (item={'name': 'efs',
'role': 'storage'})
included: create_subnet.yml for localhost => (item={'name': 'dmz',
'role': 'public'})
```

We then get the results of each of the four included task runs, the first of which looks like the following:

```
TASK [roles/subnets : Create subnet in the availability zone] ********
**********************************************
changed: [localhost] => (item={'state': 'available', 'opt_in_status':
'opt-in-not-required', 'messages': [], 'region_name': 'eu-west-1',
'zone_name': 'eu-west-1a', 'zone_id': 'euw1-az1', 'group_name':
'eu-west-1', 'network_border_group': 'eu-west-1', 'zone_type':
'availability-zone'})
changed: [localhost] => (item={'state': 'available', 'opt_in_status':
'opt-in-not-required', 'messages': [], 'region_name': 'eu-west-1',
'zone_name': 'eu-west-1b', 'zone_id': 'euw1-az2', 'group_name':
'eu-west-1', 'network_border_group': 'eu-west-1', 'zone_type':
'availability-zone'})
changed: [localhost] => (item={'state': 'available', 'opt_in_status':
'opt-in-not-required', 'messages': [], 'region_name': 'eu-west-1',
'zone_name': 'eu-west-1c', 'zone_id': 'euw1-az3', 'group_name':
'eu-west-1', 'network_border_group': 'eu-west-1', 'zone_type':
'availability-zone'})
```

As you can see, a subnet is created for each of the zones in the eu-west-1 region – this is then repeated three more times. Once the subnets have all been added, we grab more information on what has been created.

Next, the Internet Gateway role is run:

```
TASK [roles/gateway : Create an Internet Gateway] *********
changed: [localhost]
TASK [roles/gateway : Create a route table so the internet gateway can
be used by the public subnets] ****************
changed: [localhost]
```

As you may have remembered, there isn't much happening in that role, unlike the next one, which adds the network security groups, where we start by getting your current public IP address:

```
TASK [roles/securitygroups : Find out your current public IP address
using https://ipify.org/] **********************
ok: [localhost]
TASK [roles/securitygroups : Set your public ip as a fact]*
ok: [localhost]
```

As you may recall, we create the two groups in two parts – first, we create the base groups:

```
TASK [roles/securitygroups : Create the base security groups] ********
*********************************************
changed: [localhost] => (item={'name': 'learnansible-elb-security-
group', 'description': 'opens port 80 and 443 to the world', 'id_var_
name': 'elb_group_id', 'rules': [{'proto': 'tcp', 'from_port': '80',
'to_port': '80', 'cidr_ip': '0.0.0.0/0', 'rule_desc': 'allow all on
port 80'}, {'proto': 'tcp', 'from_port': '443', 'to_port': '443',
'cidr_ip': '0.0.0.0/0', 'rule_desc': 'allow all on port 443'}]})
changed: [localhost] => (item={'name': 'learnansible-ec2-security-
group', 'description': 'opens port 22 to a trusted IP and port
80 to the elb group', 'id_var_name': 'ec2_group_id', 'rules':
[{'proto': 'tcp', 'from_port': '22', 'to_port': '22', 'cidr_ip':
'86.177.22.88/32', 'rule_desc': 'allow 86.177.22.88/32 access to port
22'}, {'proto': 'tcp', 'from_port': '80', 'to_port': '80', 'group_id':
'', 'rule_desc': 'allow access to port 80 from ELB'}]})
changed: [localhost] => (item={'name': 'learnansible-rds-security-
group', 'description': 'opens port 3306 to the ec2 instances', 'id_
var_name': 'rds_group_id', 'rules': [{'proto': 'tcp', 'from_port':
'3306', 'to_port': '3306', 'group_id': '', 'rule_desc': 'allow  access
to port 3306'}]})
changed: [localhost] => (item={'name': 'learnansible-efs-security-
group', 'description': 'opens port 2049 to the ec2 instances', 'id_
var_name': 'efs_group_id', 'rules': [{'proto': 'tcp', 'from_port':
'2049', 'to_port': '2049', 'group_id': '', 'rule_desc': 'allow  access
to port 2049'}]})
```

Then we get information on the bases we have just launched and set them as facts:

```
TASK [roles/securitygroups : Set the fact for the security group ids]
*****************************************************
ok: [localhost] => (item={'name': 'learnansible-elb-security-group',
'description': 'opens port 80 and 443 to the world', 'id_var_name':
'elb_group_id', 'rules': [{'proto': 'tcp', 'from_port': '80', 'to_
port': '80', 'cidr_ip': '0.0.0.0/0', 'rule_desc': 'allow all on port
80'}, {'proto': 'tcp', 'from_port': '443', 'to_port': '443', 'cidr_
ip': '0.0.0.0/0', 'rule_desc': 'allow all on port 443'}]})
ok: [localhost] => (item={'name': 'learnansible-ec2-security-group',
'description': 'opens port 22 to a trusted IP and port 80 to the elb
group', 'id_var_name': 'ec2_group_id', 'rules': [{'proto': 'tcp',
'from_port': '22', 'to_port': '22', 'cidr_ip': '86.177.22.88/32',
'rule_desc': 'allow 86.177.22.88/32 access to port 22'}, {'proto':
```

```
'tcp', 'from_port': '80', 'to_port': '80', 'group_id': '', 'rule_
desc': 'allow access to port 80 from ELB'}]})

ok: [localhost] => (item={'name': 'learnansible-rds-security-group',
'description': 'opens port 3306 to the ec2 instances', 'id_var_name':
'rds_group_id', 'rules': [{'proto': 'tcp', 'from_port': '3306', 'to_
port': '3306', 'group_id': '', 'rule_desc': 'allow  access to port
3306'}]})

ok: [localhost] => (item={'name': 'learnansible-efs-security-group',
'description': 'opens port 2049 to the ec2 instances', 'id_var_name':
'efs_group_id', 'rules': [{'proto': 'tcp', 'from_port': '2049', 'to_
port': '2049', 'group_id': '', 'rule_desc': 'allow  access to port
2049'}]})
```

Lastly, we then add the rules; you will notice from the output that we are passing in the IDs of the groups we have created so that we can use them as part of the rules:

```
TASK [roles/securitygroups : Provision security group rules] *********
*********************************************
changed: [localhost] => (item={'name': 'learnansible-elb-security-
group', 'description': 'opens port 80 and 443 to the world', 'id_var_
name': 'elb_group_id', 'rules': [{'proto': 'tcp', 'from_port': '80',
'to_port': '80', 'cidr_ip': '0.0.0.0/0', 'rule_desc': 'allow all on
port 80'}, {'proto': 'tcp', 'from_port': '443', 'to_port': '443',
'cidr_ip': '0.0.0.0/0', 'rule_desc': 'allow all on port 443'}]})
changed: [localhost] => (item={'name': 'learnansible-ec2-security-
group', 'description': 'opens port 22 to a trusted IP and port
80 to the elb group', 'id_var_name': 'ec2_group_id', 'rules':
[{'proto': 'tcp', 'from_port': '22', 'to_port': '22', 'cidr_ip':
'86.177.22.88/32', 'rule_desc': 'allow 86.177.22.88/32 access to port
22'}, {'proto': 'tcp', 'from_port': '80', 'to_port': '80', 'group_id':
'sg-04f31e782e30e1f0a', 'rule_desc': 'allow access to port 80 from
ELB'}]})
changed: [localhost] => (item={'name': 'learnansible-rds-security-
group', 'description': 'opens port 3306 to the ec2 instances', 'id_
var_name': 'rds_group_id', 'rules': [{'proto': 'tcp', 'from_port':
'3306', 'to_port': '3306', 'group_id': 'sg-05bffd3eb96602519', 'rule_
desc': 'allow sg-05bffd3eb96602519 access to port 3306'}]})
changed: [localhost] => (item={'name': 'learnansible-efs-security-
group', 'description': 'opens port 2049 to the ec2 instances', 'id_
var_name': 'efs_group_id', 'rules': [{'proto': 'tcp', 'from_port':
'2049', 'to_port': '2049', 'group_id': 'sg-05bffd3eb96602519', 'rule_
desc': 'allow sg-05bffd3eb96602519 access to port 2049'}]})
```

Now, with the rules configured, we can start deploying some resources that use them, starting with the Target Group and ELB:

```
TASK [roles/elb : Provision the target group] *************
changed: [localhost]
TASK [roles/elb : Provision an application elastic load balancer] ****
*************************************************
changed: [localhost]
```

Then EFS:

```
TASK [roles/efs : Generate the efs targets vars file] *****
changed: [localhost]
TASK [roles/efs : Include the efs targets vars file] ******
ok: [localhost]
TASK [roles/efs : Create the EFS File System] ************
changed: [localhost]
```

Now RDS:

```
TASK [roles/rds : Add RDS subnet group] ******************
changed: [localhost]
TASK [roles/rds : Create the RDS instance] ***************
changed: [localhost]
```

Now it is time to create the temporary EC2 instance. First, we find the AMI to use:

```
TASK [roles/ec2tmp : Gather information about AMIs with the specified
filters] ****************************************
ok: [localhost]
TASK [roles/ec2tmp : filter the list of AMIs to find the latest one]
***************************************************
ok: [localhost]
```

Then, we create the SSH key pair:

```
TASK [roles/ec2tmp : Create an SSH Key Pair] *************
changed: [localhost]
```

Then, we create the EC2 instance itself:

```
TASK [roles/ec2tmp : Create the temporary ec2 instance] ***
changed: [localhost]
```

With the instance configured, we need to wait for it to have a status of **running**:

```
TASK [roles/ec2tmp : Get information about the temporary EC2 instance
to see if it is running] ***
FAILED - RETRYING: [localhost]: Get information about the temporary
EC2 instance to see if it is running (50 retries left).
. . . .
FAILED - RETRYING: [localhost]: Get information about the temporary
EC2 instance to see if it is running (46 retries left).
ok: [localhost]
```

Now that the instance is running, we add the newly launching EC2 instance to our host group:

```
TASK [roles/ec2tmp : Add the temporary EC2 instance to the vmgroup] **
***************************************************
changed: [localhost]
TASK [roles/ec2tmp : Wait for the temporary EC2 instance to be ready
to accept SSH connections] ***********************
ok: [localhost]
```

Before we move on to connecting to the EC2 host to install and configure the software stack and WordPress, we generate the endpoints variables file:

```
TASK [roles/endpoints : Generate the aws endpoints file] **
changed: [localhost]
```

That concludes the first section of the site.yml file, and we can now SSH into the temporary EC2 host and install everything:

```
PLAY [Install and configure Wordpress] ********************
TASK [Gathering Facts] ***********************************
ok: [ec2-18-203-221-2.eu-west-1.compute.amazonaws.com]
```

We then progress with the installation, which, as we have already discussed, is pretty much the same set of tasks that we covered *in Chapter 5, Deploying WordPress*, and *Chapter 9, Moving to the Cloud –* except for these tasks, which mount the EFS file system:

```
TASK [roles/stack_config : Check that the EFS volume is ready] *******
***************************************************
ok: [ec2-18-203-221-2.eu-west-1.compute.amazonaws.com]
TASK [roles/stack_config : ensure rpcbind service is running] ********
*********************************************
ok: [ec2-18-203-221-2.eu-west-1.compute.amazonaws.com]
TASK [roles/stack_config : mount the EFS volume] **********
changed: [ec2-18-203-221-2.eu-west-1.compute.amazonaws.com]
```

These tasks get the PHP version and set it as a fact:

```
TASK [roles/stack_config : Get the PHP version] ***********
changed: [ec2-18-203-221-2.eu-west-1.compute.amazonaws.com]
TASK [roles/stack_config : Set the PHP version] ***********
ok: [ec2-18-203-221-2.eu-west-1.compute.amazonaws.com]
```

Once that is complete, NGINX and PHP-FPM are restarted:

```
RUNNING HANDLER [roles/stack_config : restart nginx] ******
changed: [ec2-18-203-221-2.eu-west-1.compute.amazonaws.com]
```

```
RUNNING HANDLER [roles/stack_config : restart php-fpm] ****
changed: [ec2-18-203-221-2.eu-west-1.compute.amazonaws.com]
```

This concludes the tasks that bootstrap our temporary EC2 instance. We can now move back to our local machine and run the final section of the `sites.yml` file.

First, we create the AMI and terminate the temporary EC2 instance:

```
TASK [roles/ec2ami : find out some facts about the instance we have
been using] ****************************************
ok: [localhost]
TASK [roles/ec2ami : create the AMI] *********************
changed: [localhost]
TASK [roles/ec2ami : remove any temporary instances which are running]
**************************************************
changed: [localhost]
```

Then, we wait for two minutes:

```
TASK [roles/ec2ami : wait for 2 minutes before continuing]
Pausing for 120 seconds
(ctrl+C then 'C' = continue early, ctrl+C then 'A' = abort)
ok: [localhost]
```

Now, we grab the details of the AMI we just created:

```
TASK [roles/autoscaling : Search for all of our AMIs] *****
ok: [localhost]
TASK [roles/autoscaling : Find the last one we built] *****
ok: [localhost]
TASK [roles/autoscaling : Grab AMI ID and name of the most recent
result] ********************************************
ok: [localhost]
```

Once we have those details, we create (or if we have already run the playbook, update) the Launch Template:

```
TASK [roles/autoscaling : Create the launch template] *****
changed: [localhost]
```

Now, we gather the information needed for us to create/update the Auto Scaling Group:

```
TASK [roles/autoscaling : find out the target group ARN] **
ok: [localhost]
TASK [roles/autoscaling : get information on the ec2 subnets] *******
**********************************************
ok: [localhost]
```

Then we create the list of subnets the Auto Scaling Group will use:

```
TASK [roles/autoscaling : create a list of subnet IDs] ****
ok: [localhost] => (item={'availability_zone': 'eu-west-1c',
'availability_zone_id': 'euw1-az3', 'available_ip_address_count': 27,
'cidr_block': '10.0.0.64/27', 'default_for_az': False, 'map_public_ip_
on_launch': False, 'map_customer_owned_ip_on_launch': False, 'state':
'available', 'subnet_id': 'subnet-091ea1834c5fc8e48', 'vpc_id':
'vpc-008808ff628883751', 'owner_id': '687011238589', 'assign_ipv6_
address_on_creation': False, 'ipv6_cidr_block_association_set': [],
'tags': {'role': 'compute', 'deployedBy': 'Ansible', 'Name': 'ec2-
subnet-euw1-az3', 'environment': 'prod', 'description': 'Resource
managed by Ansible', 'projectName': 'learnansible'}, 'subnet_arn':
'arn:aws:ec2:eu-west-1:687011238589:subnet/subnet-091ea1834c5fc8e48',
'enable_dns64': False, 'ipv6_native': False, 'private_dns_name_
options_on_launch': {'hostname_type': 'ip-name', 'enable_resource_
name_dns_a_record': False, 'enable_resource_name_dns_aaaa_record':
False}, 'id': 'subnet-091ea1834c5fc8e48'})
```

The preceding output is repeated twice for the other two subnets we will be using; then, we finally create/update the Auto Scaling Group:

```
TASK [roles/autoscaling : Create/update the auto-scaling group using
the launch template we just created] **********
changed: [localhost]
```

Now, we have come to the end of our Playbook run, and we get the recap:

```
PLAY RECAP ***************************************************
ec2-18-203-221-2.eu-west-1.compute.amazonaws.com :
ok=37    changed=28    unreachable=0    failed=0    skipped=1    res-
cued=0    ignored=2
localhost :
ok=56    changed=23    unreachable=0    failed=0    skipped=30    res-
cued=0    ignored=0
```

When I ran the playbook, it took just over 20 minutes to complete the first time, with subsequent runs taking around 10 minutes to finish.

So, from a single command and in 20ish minutes, we have a highly available vanilla WordPress installation. If you find out the public URL of your Elastic Load Balancer from the AWS console or by checking the value of the elb key in the group_vars/generated_aws_endpoints.yml file, you should be able to see your site.

## Terminating all the resources

Before we complete this chapter, we need to look at terminating the resources; to do this, you can run the following:

```
$ ansible-playbook -i hosts destroy.yml
```

This removes everything in the reverse order that we launched it, starting with the Auto Scaling Group:

```
PLAY [Destroy the AWS Environment created by the site.yml playbook]
************
TASK [Gathering Facts] *********************************
ok: [localhost]
TASK [Delete the Auto Scaling Group] *********************
changed: [localhost]
TASK [Delete the Launch Template] ************************
changed: [localhost]
```

As there can be more than one AMI, we gather some facts and then loop through removing everything that is returned:

```
TASK [Get information about the AMIs] ********************
ok: [localhost]
TASK [Delete the AMI(s)] ********************************
changed: [localhost] => (item={'architecture': 'x86_64',
'creation_date': '2024-01-12T09:44:07.000Z', 'image_id': 'ami-
0ddfeb5a1fb64c23a', 'image_location': '687011238589/learnansible-
prod-ami-2024-01-12_0944', 'image_type': 'machine', 'public':
False, 'tags': {'Name': 'learnansible-prod-ami-2024-01-12_0944',
'deployedBy': 'Ansible', 'environment': 'prod', 'buildDate': '2024-
01-12 09:44:06', 'description': 'Resource managed by Ansible',
'projectName': 'learnansible', 'role': 'ami'}, 'virtualization_type':
'hvm', 'source_instance_id': 'i-050689909fa289998'})
```

We then remove more one-off resources:

```
TASK [Create a SSH Key Pair] ***************************
changed: [localhost]
TASK [Delete the group_vars/generated_aws_endpoints.yml file] ********
************************************************
changed: [localhost]
TASK [Delete the RDS database] **************************
changed: [localhost]
TASK [Delete RDS subnet group] **************************
changed: [localhost]
TASK [Delete the group_vars/generated_rds_passwordfile file] *********
************************************************
changed: [localhost]
TASK [Delete the EFS File System] ***********************
changed: [localhost]
TASK [Delete the group_vars/generated_efs_targets.yml file]
changed: [localhost]
TASK [Delete the application elastic load balancer]********
changed: [localhost]
```

```
TASK [Delete the target group] ****************************************
******************************
changed: [localhost]
```

As the security groups reference each other, we need to create a list of them in reverse order so we can attempt to delete a group that is referenced by the next one we are going to delete:

```
TASK [Create a reversed list of the security group names] *
ok: [localhost]
TASK [Delete the security groups] *************************
changed: [localhost] => (item=learnansible-efs-security-group)
changed: [localhost] => (item=learnansible-rds-security-group)
changed: [localhost] => (item=learnansible-ec2-security-group)
FAILED - RETRYING: [localhost]: Delete the security groups (50 retries
left).
. . . . .
FAILED - RETRYING: [localhost]: Delete the security groups (46 retries
left).
changed: [localhost] => (item=learnansible-elb-security-group)
```

You may have noticed that it failed towards the end; that is because the AWS API is having a little trouble keeping up, and the playbook is running a little ahead of the results it is returning.

We check a few more tasks:

```
TASK [Get information about the VPC] ***********************
ok: [localhost]
TASK [Get information about the Route Table] **************
ok: [localhost]
TASK [Delete the Route Table] *****************************
changed: [localhost] => (item={'associations': [{'main': False,
'route_table_association_id': 'rtbassoc-0738bb9e5aaf44848',
'route_table_id': 'rtb-04bc7177949ad2c92', 'subnet_id': 'subnet-
07c28d376283741f6', 'association_state'
TASK [Delete the Internet Gateway]************************
changed: [localhost]
```

Next, we have the subnets:

```
TASK [Get information on the subnets] ********************************
******************************
ok: [localhost]

TASK [Delete the subnets] *******************************
changed: [localhost] => (item={'availability_zone': 'eu-west-1c',
'availability_zone_id': 'euw1-az3', 'available_ip_address_count': 27,
```

```
'cidr_block': '10.0.0.64/27', 'default_for_az': False, 'map_public_ip_
on_launch': False, 'map_customer_owned_ip_on_launch': False, 'state':
'available', 'subnet_id': 'subnet-091ea1834c5fc8e48', 'vpc_id': 'vpc-
008808ff628883751', 'id': 'subnet-091ea1834c5fc8e48'})

. . . . .

changed: [localhost] => (item={'availability_zone': 'eu-west-1b',
'availability_zone_id': 'euw1-az2', 'available_ip_address_count':
27, 'cidr_block': '10.0.0.128/27', 'default_for_az': False, 'map_
public_ip_on_launch': False, 'map_customer_owned_ip_on_launch': False,
'state': 'available', 'subnet_id': 'subnet-0fd4610392872d442', 'vpc_
id': 'vpc-008808ff628883751', 'id': 'subnet-0fd4610392872d442'})
```

Finally, we get to the VPC and recap:

```
TASK [Delete the VPC] ************************************
changed: [localhost]
PLAY RECAP ****************************************************
localhost :
ok=23    changed=17    unreachable=0    failed=0    skipped=0    res-
cued=0    ignored=0
```

Once the playbook has finished running, I recommend you log in to the AWS console and double-check that everything has been correctly removed, as you don't want to incur any unexpected costs.

## Summary

In this chapter, we have taken our AWS deployment to the next level by creating and launching a highly available WordPress installation. By leveraging the various services offered by AWS, we engineered out any single points of failure regarding the availability of instances and our use of availability zones.

We also built logic into our playbook to use the same command to launch a new deployment or update the operating system on an existing one with a rolling deployment of new instance AMIs that contain our updated packages, leading to zero downtime during deployment.

While the WordPress deployment is as simple as possible, deploying the production-ready images would remain similar when using a more complicated application.

In our next chapter, we will look at moving from the public to the private cloud and how Ansible interacts with VMware.

# 12

# Building Out a VMware Deployment

Now that we know how to launch networking and services in AWS, we will discuss deploying a similar setup in a **VMware** environment and talk through the core VMware modules.

In this chapter, we will cover the following topics:

- An introduction to VMware
- The VMware REST modules

## Technical requirements

This chapter will discuss various components of the VMware family of products and how you can interact with them using Ansible. While there will be example playbook tasks in this chapter, they may need to be more easily transferable to your installation. Because of this, it's not recommended that you use any examples in this chapter without first reviewing the complete documentation.

## An introduction to VMware

With over 25 years of history, VMware has evolved significantly from its origins as a stealth startup. Boasting a revenue of over $13 billion in August 2023, the Vmware product portfolio, which grew to encompass around 30 products, is best known for its hypervisors, and it is a staple in most enterprises, enabling administrators to deploy virtual machines rapidly across various standard x86-based hardware configurations.

However, recent developments have seen significant changes following Broadcom's acquisition of Vmware in late 2023.

This acquisition has dramatically simplified Vmware's product portfolio, something that was influenced by customer and partner feedback, allowing users of all sizes to derive more value from VMware solutions. Two notable offerings include VMware Cloud Foundation and VMware vSphere Foundation, each with advanced add-on offers.

The first of the major changes that Broadcom has implemented is transitioning VMware to a subscription-based model. This aligns with the industry standard for cloud consumption and aims to provide continuous innovation, quicker time to value, and predictable investments for customers by phasing out perpetual licenses and replacing them with subscription or term licenses to enable customer and partner success in digital transformations.

There are concerns from the wider industry about Broadcom's post-acquisition strategy for VMware. There's speculation that Broadcom may focus on retaining only the largest and most profitable VMware customers and partners. This strategy could lead to a restructuring of VMware's portfolio to better align with Broadcom's business objectives, potentially including asset disposals and an even more streamlined product range.

At the time of writing (early 2024), the impact of these changes on VMware's existing customer base and partner ecosystem is still unknown, with further details expected to emerge throughout the year as Broadcom continues to implement its strategic long-term plans for VMware.

## The VMware REST modules

As already mentioned, there were around 30 products in the VMware range, and Ansible had modules that allowed you to interact with many of them.

However, due to product streamlining, we will just concentrate on the `vmware.vmware_rest` namespace modules and won't be looking at any of the `community.vmware` modules as these will lose all support at some point in 2025.

The difference between the two collections of modules is that, as implied by the name, the `vmware.vmware_rest` modules use the VMware REST API to manage resources, whereas the ones in `community.vmware` use a Python library to interact with the various VMware endpoints to perform tasks.

The modules in the `vmware.vmware_rest` namespace are split into three areas:

- **Appliance**: These modules manage your vCenter appliances, which are underlying resources that make up your vCenter deployment
- **Content**: The Content Library modules allow you to manage the services for defining and managing the library's items, subscription, publication, and storage
- **vCenter**: These modules allow you to manage the workloads, such as virtual machines, running on top of your vCentre deployment

Let's start by looking at the VMware REST appliance modules.

## VMware REST appliance modules

At the time of writing, there are over 60 modules; these are split up into their own clearly labeled areas.

### Access modules

To start with, we have the access modules:

- `appliance_access_consolecli`: This module allows you to enable or disable the console-based controlled CLI (TTY1).

- `appliance_access_consolecli_info`: This module returns the current state of the console-based controlled CLI (TTY1); this will either be enabled or disabled.

- `appliance_access_dcui`: With this module, you can configure the state of the **Direct Console User Interface** (**DCUI** TTY2); again, you only have two options: enabled or disabled.

- `appliance_access_dcui_info`: As you may have already guessed, this module returns either enabled or disabled for the DCUI TTY2 state.

- `appliance_access_shell`: Again, there isn't much to this one in that you just change the enabled state of BASH. With this enabled, you will be able to access a BASH shell within the CLI.

- `appliance_access_shell_info`: This module simply returns BASH access; this will either be enabled or disabled.

- `appliance_access_ssh`: This module sets the enabled state of the SSH-based controlled CLI.

- `appliance_access_ssh_info`: This module returns the enabled state of the SSH-based controlled CLI.

As already mentioned, each of these modules either allows you to set the state of the access system or returns the currently configured state:

```
- name: "Enable SSH access"
  vmware.vmware_rest.appliance_access_ssh:
    enabled: true
  register: access_ssh_result
```

Each of the non-info modules has a single value of `enabled`, which accepts either `true` or `false`, as demonstrated earlier.

### Health info modules

The next grouping of modules only returns information about the health of your system:

- `appliance_health_applmgmt_info`
- `appliance_health_database_info`

- `appliance_health_databasestorage_info`

- `appliance_health_load_info`

- `appliance_health_mem_info`

- `appliance_health_softwarepackages_info`

- `appliance_health_storage_info`

- `appliance_health_swap_info`

- `appliance_health_system_info`

You would call one of the modules like this:

```
- name: "Get the system health status"
  vmware.vmware_rest.appliance_health_system_info:
  register: health_system_result
```

This would return the current health of whichever of the services you are querying.

### Infraprofile modules

Here, we have just two modules:

- `appliance_infraprofile_configs`: This module exports the selected profile

- `appliance_infraprofile_configs_info`: This module lists all the registered profiles

The only valid state for the `appliance_infraprofile_configs` module is `export`:

```
- name: "Export the ApplianceManagement profile"
  vmware.vmware_rest.appliance_infraprofile_configs:
    state: "export"
    profiles:
    - "ApplianceManagement"
  register: infraprofile_configs_result
```

Here's the output is JSON containing the profile for the selected configuration. In the preceding example, this is `ApplianceManagement`.

### Local accounts modules

Here, we have three modules:

- `appliance_localaccounts_globalpolicy`

- `appliance_localaccounts_globalpolicy_info`

- `appliance_localaccounts_info`

These modules allow you to set and query the global policy and return information on all or just one of the local accounts.

## Monitoring modules

While there are only two modules here, they can be powerful when you combine them:

- `appliance_monitoring_info`: This module returns a list of monitors

- `appliance_monitoring_query`: This module allows you to query the monitors

Here's an example query:

```
- name: "Query the monitoring backend"
  vmware.vmware_rest.appliance_monitoring_query:
    start_time: "2024-01-01 09:00:00+00:00"
    end_time: "2024-01-01 10:00:00+00:00"
    names:
    - "mem.total"
    interval: "MINUTES5"
    function: "AVG"
  register: mem_total_result
```

As you can see, with the preceding task, we are querying the total memory in 5 minutes, which averages between 9 A.M. and 10 A.M. on January 1, 2024.

## Networking modules

This is where things start to get a little more complicated; each of the modules has an *info* equivalent where highlighted:

- `appliance_networking` (plus info): This module resets and restarts network configuration on all interfaces. It also renews the DHCP leases for DHCP IP addresses.

- `appliance_networking_dns_domains` (plus info): This module is used to manage the DNS search domains.

- `appliance_networking_dns_hostname` (plus info): This module configures the **fully qualified domain name (FQDN)** hostname.

- `appliance_networking_dns_servers` (plus info): This module can manage the DNS server configuration.

- `appliance_networking_firewall_inbound` (plus info): This module sets an ordered list of firewall rules.

- `appliance_networking_interfaces_info`: This module fetches information on a single network interface.

- `appliance_networking_interfaces_ipv4` (plus info): This module manages the IPv4 network configuration for the named network interface.

- `appliance_networking_interfaces_ipv6` (plus info): This module manages the IPv6 network configuration for the named network interface.

- `appliance_networking_noproxy` (plus info): This module configures servers for which no proxy configuration should be applied.

- `Appliance_networking_proxy` (plus info): This module configures which proxy server to use for the specified protocol.

### The time and date modules

The following modules affect the time and date settings in some way:

- `appliance_ntp` (plus info): This module manages the NTP server configuration

- `appliance_system_time_info`: This module gets the system time

- `appliance_system_time_timezone` (plus info): This module sets the time zone

- `appliance_timesync module` (plus info): This module configures time sync mode

### The remaining modules

The remaining modules cover appliance configuration and management:

- `appliance_services` (plus info): You can use this module to restart a given service

- `appliance_shutdown` (plus info): This module allows you to cancel a pending shutdown action

- `appliance_system_globalfips` (plus info): Using this module, you can enable or disable Global FIPS mode for the appliance

- `appliance_system_storage` (plus info): This module resizes all partitions to 100% of the disk size

- `appliance_system_version_info`: This module gets version information

- `appliance_update_info`: This module gets the status of an appliance update

- `appliance_vmon_service` (plus info): This module lists details of services managed by vmon

This concludes the appliance section. Next, we'll look at content modules.

## VMware REST content modules

There are a small number of modules that allow you to manage and gather information on your content libraries:

- `content_configuration` (plus info): This module updates the configuration
- `content_library_item_info`: This module returns {`@link ItemModel`} when provided with an identifier
- `content_locallibrary` (plus info): This module creates a new local library
- `content_subscribedlibrary` (plus info): This module creates a new subscription

## vCenter modules

This is where the more interesting things happen. Using these modules, you can launch, configure, and manage the entire life cycle of your virtual machines. Before we look at virtual machines, we'll take a look at some of the supporting vCenter modules.

### Supporting vCenter modules

These supporting modules allow you to manage things such as data centers, folders, data stores, and resource pools hosted within your vCenter:

- `vcenter_cluster_info`: This module retrieves information about the cluster corresponding to {`@param.name cluster_name`}
- `vcenter_datacenter` (plus info): This module adds a new data center to your vCenter inventory
- `vcenter_datastore_info`: This module fetches information about the data store using {`@param.name datastore_name`}
- `vcenter_folder_info`: This module retrieves information on up to 1,000 folders in vCenter matching {`@link FilterSpec`} that the user you are connecting as has permission to see
- `vcenter_host` (plus info): This module can be used to add a new standalone host to your vCenter
- `vcenter_network_info`: This module returns information about the first 1,000 visible networks in vCenter matching {`@link FilterSpec`}, depending on your permissions
- `vcenter_ovf_libraryitem`: This module is used to create an item in the content library from a virtual machine or virtual appliance
- `vcenter_resourcepool` (plus info): This module deploys a resource pool
- `vcenter_storage_policies_info`: This module fetches information about the storage policies available in vCenter; it returns a maximum of 1,024 results

### Virtual machine modules

The final group of modules deals with creating and managing virtual machines and their associated resources. Let's start by looking at the main module, vcenter_vm.

The vcenter_vm module is used to create virtual machines. For example, a basic task would look like this:

```
- name: "Create a Virtual Machine"
  vmware.vmware_rest.vcenter_vm:
    placement:
      cluster: "{{ lookup('vmware.vmware_rest.cluster_moid', '/
  learnansible_dc/host/learnansible_cluster') }}"

    folder: "{{ lookup('vmware.vmware_rest.folder_moid', '/learnansible_
  dc/vm') }}"
      resource_pool: "{{ lookup('vmware.vmware_rest.resource_pool_
  moid', '/learnansible_dc/host/learnansible_cluster/Resources') }}"
    name: "LearnAnsibleVM"
    guest_OS: "UBUNTU_64"
    hardware_version: "VMX_11"
    memory:
      hot_add_enabled: true
      size_MiB: 4000
  register: LearnAnsibleVM_output
```

As you can see, we are using a few of the different lookup modules to find the cluster, data store, folder, and resource pool IDs – if we had this information, we could provide the IDs directly.

Once the virtual machine has been created, we can use the remaining modules to configure it more or manage its state:

- vcenter_vm_guest_customization: This module applies guest customization to the virtual machine, such as running a script

- vcenter_vm_guest_filesystem_directories: Using this module, you can create a directory within the guest host operating system

- vcenter_vm_guest_identity_info: This module fetches information about the guest host

- vcenter_vm_guest_localfilesystem_info: This module grabs details of the local filesystems in the guest host operating system

- vcenter_vm_guest_networking_info: This module fetches details about the network configuration within the guest host operating system

- `vcenter_vm_guest_networking_interfaces_info`: This module displays information about the network interfaces in the guest host operating system

- `vcenter_vm_guest_networking_routes_info`: This module displays information about the network routes from within the guest host operating system

- `vcenter_vm_guest_operations_info`: This module grabs information about the guest host operating system's status

- `vcenter_vm_guest_power` (plus info): This module requests a soft shutdown, standby (suspend), or soft reboot from within the guest host operating system

- `vcenter_vm_hardware`: This module is used to update the hardware settings of the requested virtual machine

- `vcenter_vm_hardware_adapter_sata` (plus info): This module configures a virtual SATA adapter

- `vcenter_vm_hardware_adapter_scsi` (plus info): This module adds a virtual SCSI adapter

- `vcenter_vm_hardware_boot` (plus info): This module is used to manage virtual machine boot-related settings

- `vcenter_vm_hardware_boot_device` (plus info): This module can set the virtual devices that will be used as the boot drive for your virtual machine

- `vcenter_vm_hardware_cdrom` (plus info): This module attaches a virtual CD-ROM to your virtual machine

- `vcenter_vm_hardware_cpu` (plus info): This module manages your virtual machine's CPU settings

- `vcenter_vm_hardware_disk` (plus info): This module connects virtual disks to your virtual machine

- `vcenter_vm_hardware_ethernet` (plus info): This module connects a virtual Ethernet adapter to your virtual machine

- `vcenter_vm_hardware_floppy` (plus info): This module adds a virtual floppy drive to the virtual machine

- `vcenter_vm_hardware_info`: This module fetches your virtual machine's virtual hardware settings information

- `vcenter_vm_hardware_memory` (plus info): This module configures the memory settings

- `vcenter_vm_hardware_parallel` (plus info): This module adds a virtual parallel port

- `vcenter_vm_hardware_serial` (plus info): This module adds a virtual serial port

- `vcenter_vm_info`: This module returns information about your virtual machine

- `vcenter_vm_libraryitem_info`: This module retrieves information about the library item associated with your virtual machine

- `vcenter_vm_power` (plus info): This module issues a boot, hard shutdown, hard reset, or hard suspend on a guest – that is, it presses the power button on the front

- `vcenter_vm_storage_policy` (plus info): This module updates the storage policy of your virtual machine's virtual hard disks

- `vcenter_vm_storage_policy_compliance`: This module updates and gathers information on your virtual machine's storage policy compliance

- `vcenter_vm_tools` (plus info): This module is used to manage the configuration of VMware Tools

- `vcenter_vm_tools_installer` (plus info): This module attaches the VMware Tools CD installer as a CD-ROM, making it available within the guest host operating system

- `vcenter_vmtemplate_libraryitems` (plus info): This module creates and returns information on items in the content library

As you can see, there is comprehensive support for managing your virtual machine resources using the `vmware.vmware_rest` collection, and what's better is that the modules are all designed to consume the official REST API, meaning that you can safely mix and match how you manage your resources within VMware, regardless of whether you use the CLI, web interface, or Ansible. Everything is managed via the same REST API.

## Summary

As you have seen from the very long list of modules, you can do most of the management and configuration tasks you would be doing as a VMware administrator day-to-day using Ansible.

Add to this the modules we looked at in *Chapter 8, Ansible Network Modules,* for managing network equipment, and modules such as the ones that support hardware such as NetApp storage devices.

By doing this, you can build complex playbooks that span the physical devices, VMware elements, and virtual machines running within your on-premises enterprise-level virtualized infrastructure.

As mentioned at the start of this chapter, at the time of writing, there is a lot of upheaval at VMware. This chapter has been written to show the art of the possible rather than be a practical hands-on guide for managing your VMware resources using Ansible. For more details on the current state of the `vmware.vmware_rest` collection, go to `https://galaxy.ansible.com/ui/repo/published/vmware/vmware_rest/`.

In the next chapter, we will look at how to ensure that our playbooks are following best practices by scanning them for common issues and potential security problems.

# Part 4: Ansible Workflows

In the final part of this book, you will learn advanced Ansible workflows and best practices, including security practices, playbook scanning, server hardening, CI/CD integration, Ansible AWX, and Red Hat Ansible Automation Platform. By the end, you will be equipped with the knowledge you need to effectively utilize Ansible in real-world scenarios.

This part has the following chapters:

- *Chapter 13, Scanning Your Ansible Playbooks*
- *Chapter 14, Hardening Your Servers Using Ansible*
- *Chapter 15, Using Ansible with GitHub Actions and Azure DevOps*
- *Chapter 16, Introducing Ansible AWX and Red Hat Ansible Automation Platform*
- *Chapter 17, Next Steps with Ansible*

# 13

# Scanning Your Ansible Playbooks

In this chapter, you will learn how to scan your Ansible playbooks using two third-party tools: Checkov and KICS. Both are open source and can help you identify and fix common configuration issues within your Ansible code, such as syntax errors, misconfigurations, hardcoded secrets, and deployment problems, which could lead to potential breaches.

By the end of this chapter, you will have done the following:

- Installed and run Checkov and KICS scans on our Ansible playbooks
- Reviewed the results and reports generated during the scans
- Fixed any issues detected during the scans

The chapter covers the following topics:

- Why scan your playbooks?
- Docker overview and installation
- Exploring Checkov
- Exploring KICS

## Technical requirements

Rather than installing the tools locally, we will use Docker to execute the scans; there will be a little detail on how to install Docker later in the chapter. Additionally, we will be scanning a variation of the playbook we wrote in *Chapter 11, Highly Available Cloud Deployments*; this can be found in the repository at `https://github.com/PacktPublishing/Learn-Ansible-Second-Edition/tree/main/Chapter13`.

# Why scan your playbooks?

While we have been taking a sensible approach to deploying our cloud resources in previous chapters, many of the guardrails we have put in place have all been ones I have learned through experience and by applying a little common sense.

For example, when launching a virtual machine resource in either Microsoft Azure or Amazon Web Services, we have been locking down the SSH or RDP service to the host's public IP address, which is running Ansible; up until now, this has been your local machine rather than just opening SSH or RDP to the world by using `0.0.0.0/0` as the source address, which is the CIDR notation for *"allow all."*

This is not a problem for the workloads we have been working on; having a virtual machine exposed directly to the internet with its management port open for everyone to access is not considered best practice, as it will expose you to brute-force attacks, which, if they are successful, will not only lead to that machine being compromised; it could also act as a gateway to the rest of your network and other associated resources such as databases and storage.

I would class the preceding example as common sense, but as we launch more and more cloud services using our playbooks, how can we ensure that we are following best practices for services that maybe we haven't had much experience with outside of getting something up and running? How can we put some guardrails in place to stop us from doing something before resources are deployed?

This is where the two tools we will look at in this chapter come in; they are designed to scan your playbooks, look at the configuration, and compare them to their best practice policies. Eventually, in *Chapter 15, Using Ansible with GitHub Actions and Azure DevOps*, we will build one of the two tools into our deployment pipelines, but for now, we are going to look at the tools and run them locally using Docker.

# Docker overview and installation

Docker, the platform that made containers popular, is both an open source and commercial solution that enables you to package all of the elements of your application, including libraries and other dependencies, alongside your own code in a single, easy-to-distribute package; this means that we won't need to download and install all of the prerequisites for the tools that we will be running in this chapter or need to compile the tools from source to get working executables for our system.

To follow the example in this chapter, you must install **Docker Desktop** on your host.

## Installing Docker Desktop on macOS

To install Docker Desktop on macOS, follow these three steps:

1.  Choose the appropriate installer for your Mac's architecture:

    I.   For ARM64 (Apple Silicon), use `https://desktop.docker.com/mac/main/arm64/Docker.dmg`.

II. For AMD64 (Intel Macs), use `https://desktop.docker.com/mac/main/amd64/Docker.dmg`.

2. After downloading, open the `Docker.dmg` file by double-clicking it. In the opened window, drag the Docker icon to your Applications folder to install Docker Desktop. It will be installed at `/Applications/Docker.app`.

3. To launch Docker, navigate to the Applications folder and double-click on **Docker**; this will start **Docker Desktop**.

When you first launch Docker Desktop, it will walk you through the remaining installation steps and run in the background once complete.

## Installing Docker Desktop on Windows

To install Docker Desktop on Windows, follow these instructions:

1. Download the Docker Desktop Installer for Windows from this link: `https://desktop.docker.com/win/main/amd64/Docker%20Desktop%20Installer.exe`.

2. Run the downloaded **Docker Desktop Installer.exe** by double-clicking on it. Docker Desktop will be installed at the default location `C:\Program Files\Docker\Docker`.

3. During the installation, you may be prompted to choose whether to use **WSL 2 (Windows Subsystem for Linux 2)** or **Hyper-V** as the backend. Select the **Use WSL 2 instead of Hyper-V** option, as we have used this throughout the book to run Ansible.

4. Follow the on-screen instructions provided by the installation wizard to authorize the installer and complete the installation process.

5. Once the installation is completed, click **Close** to finish the setup.

From here, you can open Docker Desktop from the start menu, and it will run in the background.

## Installing Docker Desktop on Linux

If you are running a Linux Desktop, the instructions will differ slightly depending on your Linux distribution; for detailed instructions, see `https://docs.docker.com/desktop/linux/install/`.

Now, with Docker Desktop installed, we can look at the first of the two tools we will look at.

# Exploring Checkov

Checkov is an open source static code analysis tool maintained by Prisma Cloud designed for **infrastructure-as-code (IaC)**.

It helps developers and DevOps teams identify misconfigurations in their files before deployment to cloud environments. By scanning the code for tools such as Terraform, CloudFormation, Kubernetes, and others, including Ansible, Checkov checks against best practices and compliance guidelines, ensuring your infrastructure deployments are secure, efficient, and compliant with industry standards before it is deployed.

> **Important note**
>
> You may have noticed that Ansible is mentioned as "others" in the preceding description; that is because support for Ansible was only just introduced at the time of writing this in early 2024. Because of this, while we will be looking at Checkov during this chapter, we will not be going into as much detail about Checkov or the second tool, Kics.

Before we run our scan, we need a playbook; open your terminal and check out the scan GitHub repository by running the following:

```
$ git clone git@github.com:PacktPublishing/Learn-Ansible-Second-
Edition.git
```

This repository contains a copy of the final playbook code from *Chapter 11*, *Highly Available Cloud Deployments*.

Now that we have the code checked out, we can download the Checkov container image. To do this, we need to pull it from Docker Hub by running the following command:

```
$ docker image pull bridgecrew/checkov:latest
```

This will download the image from `https://hub.docker.com/r/bridgecrew/checkov`, and with it downloaded, we can now scan our playbook code.

To run the scan, issue the following commands:

```
$ cd Learn-Ansible-Second-Edition/Chapter13
$ docker container run --rm --tty --volume ./:/ansible --workdir /
ansible bridgecrew/checkov --directory /ansible
```

Before we review the results, let's quickly break down the command that we have just run:

- `docker container run` executes a new Docker container.
- `--rm` instructs Docker to remove the container after it exits automatically.
- `--tty` allocates a pseudo-TTY, which makes the scan output readable to our session.
- `--volume ./:/ansible` mounts the current directory, defined as `./`, to the `/ansible` path inside the container.
- `--workdir /ansible` sets the working directory inside the container to `/ansible`.

- `bridgecrew/checkov` specifies the Checkov Docker image we have just pulled from the Docker Hub.

- `--directory /ansible` instructs Checkov to scan files in `/ansible`; it is not part of the Docker command but is sending instructions to the Checkov binary, which is the default entry point for our container to run the scan. If we had Checkov installed locally, then this would be the equivalent to running the `checkov --directory /ansible` command.

Now that we have broken down the command used to run the scan, we can look at the output of the scan itself, starting with the overview:

```
ansible scan results:
Passed checks: 5, Failed checks: 3, Skipped checks: 0
```

As you can see, we have more passes than failed checks, which is a good start; the next section of the output details the checks, starting with the following passes:

```
Check: CKV_ANSIBLE_2: "Ensure that certificate validation isn't
disabled with get_url"
        PASSED for resource: tasks.ansible.builtin.get_url.download wp-
cli
        File: /roles/wordpress/tasks/main.yml:11-17
```

Our first pass checks to see if we are instructing the `ansible.builtin.get_url` module to bypass certificate validation when connecting to an HTTPS site to download content.

The next four passes are for the two times our playbook uses the `ansible.builtin.apt` module:

```
Check: CKV_ANSIBLE_5: "Ensure that packages with untrusted or missing
signatures are not used"
        PASSED for resource: tasks.ansible.builtin.apt.update apt-cache
and upgrade packages
        File: /roles/stack-install/tasks/main.yml:5-13
Check: CKV_ANSIBLE_5: "Ensure that packages with untrusted or missing
signatures are not used"
        PASSED for resource: tasks.ansible.builtin.apt.update cache and
install the stack packages
        File: /roles/stack-install/tasks/main.yml:27-33
```

The first pair of passes ensures that we are not installing any packages that are not correctly signed. The second pair of passes also checks for the same thing:

```
Check: CKV_ANSIBLE_6: "Ensure that the force parameter is not used, as
it disables signature validation and allows packages to be downgraded
which can leave the system in a broken or inconsistent state"
        PASSED for resource: tasks.ansible.builtin.apt.update apt-cache
and upgrade packages
        File: /roles/stack-install/tasks/main.yml:5-13
```

```
Check: CKV_ANSIBLE_6: "Ensure that the force parameter is not used, as
it disables signature validation and allows packages to be downgraded
which can leave the system in a broken or inconsistent state"
      PASSED for resource: tasks.ansible.builtin.apt.update cache and
install the stack packages
      File: /roles/stack-install/tasks/main.yml:27-33
```

However, this time, the check ensures that we are not using the force parameter, which, as you can see from the description, disables signature checks and can also leave our APT database in a little bit of a state if things go wrong.

Now, we move on to the failures; the first failure is the one we called out as the example when we spoke about why you would want to use the tools we are covering in this chapter:

```
Check: CKV_AWS_88: "EC2 instance should not have public IP."
      FAILED for resource: tasks.amazon.aws.ec2_instance.Create the
temporary ec2 instance
      File: /roles/ec2tmp/tasks/main.yml:53-75
      Guide: https://docs.prismacloud.io/en/enterprise-edition/policy-
reference/aws-policies/public-policies/public-12
            62 |      network:
            63 |          assign_public_ip: "{{ ec2.public_ip }}"
```

So, what gives? As you may recall from *Chapter 11*, *Highly Available Cloud Deployments*, the instance we are launching is only temporary and accessible while the playbook is running. However, Checkov doesn't know this, so it rightly calls it out and, as you can see, provides details on why this is via the guide URL, which, for this check, is https://docs.prismacloud.io/en/enterprise-edition/policy-reference/aws-policies/public-policies/public-12.

Moving on to the next failure in the scan, we see the following:

```
Check: CKV_AWS_135: "Ensure that EC2 is EBS optimized"
      FAILED for resource: tasks.amazon.aws.ec2_instance.Create the
temporary ec2 instance
      File: /roles/ec2tmp/tasks/main.yml:53-75
      Guide: https://docs.prismacloud.io/en/enterprise-edition/policy-
reference/aws-policies/aws-general-policies/ensure-that-ec2-is-ebs-
optimized
```

In this case, Checkov believes a parameter is missing from the amazon.aws.ec2_instance block when we launch the temporary EC2 instance. It is recommended that the parameter ebs_optimized is set to true rather than keeping the value as false, which is the default for the parameter.

The final failure in the scan output is as follows:

```
Check: CKV2_ANSIBLE_2: "Ensure that HTTPS url is used with get_url"
      FAILED for resource: tasks.ansible.builtin.get_url.download wp-
cli
```

```
File: /roles/wordpress/tasks/main.yml:11-17
        11 |  - name: "download wp-cli"
        12 |    ansible.builtin.get_url:
        13 |      url: "{{ wp_cli.download }}"
        14 |      dest: "{{ wp_cli.path }}"
```

As Checkov is doing static code analysis, it isn't designed to check for the contents of variables. Because the policy checks that we are providing a secure URL (that is, `https://domain.com/` in the `url` section of the task) it fails, as it is just seeing the `{{ wp_cli.download }}` variable name rather than the contents of the variable.

If you are keeping count, that makes two of the three failed checks false positives; for the first failure, we can accept the risk, as we know the machine is only temporary, and we know that we are locking the EC2 instance down to trusted IP addresses.

For the third failure, we can confirm that the contents of the `{{ wp_cli.download }}` variable is a secure URL, as it is `https://raw.githubusercontent.com/wp-cli/builds/gh-pages/phar/wp-cli.phar`.

The second failure is the only one we need to look at; let's take a look at the tasks, starting with the `Amazon.aws.ec2_instance` one.

Here, we need to add two things; the first thing is a comment to instruct Checkov that we accept the risk being highlighted by the `CKV_AWS_88` policy, and then we need to set `ebs_optimized` to `true`.

The following code shows the updates I have made to `roles/ec2tmp/tasks/main.yml`; everything below the `name` parameter remains as is:

```
- name: "Create the temporary ec2 instance"
  amazon.aws.ec2_instance:
    # checkov:skip=CKV_AWS_88:"While a public IP address is assigned
to the instance, it is locked down by the security group and the
instance is temporary."
    ebs_optimized: true
    name: "{{ ec2_tmp_name }}"
```

As you can see, instructing Checkov to skip a check is straightforward; the comment is split into four parts:

- `#` is the standard syntax for starting a comment in a YAML file

- `checkov:` instructs Checkov to pay attention to the contents of the comment

- `skip=CKV_AWS_88:` instructs Checkov to skip the `CKV_AWS_88` check when it runs

- `"While a public IP address is assigned to the instance, it is locked down by the security group and the instance is temporary."` is the suppress comment that will appear in the output when we run the scan

The next line in the update task implements the recommendation that we set the `ebs_optimized` parameter to `true`.

Now, we move on to the second task, which we need to update, and can be found in `roles/wordpress/tasks/main.yml`. Here, we just add a comment to make Checkov skip CKV2_ANSIBLE_2:

```
- name: "download wp-cli"
  ansible.builtin.get_url:
    # checkov:skip=CKV2_ANSIBLE_2:"The URL passed in the variable is
secured with SSL/TLS protocol."
    url: "{{ wp_cli.download }}"
    dest: "{{ wp_cli.path }}"
```

If you are following along, the repository contains a branch called `checkov`; with the preceding detailed changes applied, you can switch to it by running the following:

```
$ git switch chapter13-checkov
```

Then, we can re-run the scan using the following command:

```
$ docker container run --rm --tty --volume ./:/ansible --workdir /
ansible bridgecrew/checkov --directory /ansible
```

I can see that my changes have both suppressed and resolved the three failures:

```
ansible scan results:
Passed checks: 6, Failed checks: 0, Skipped checks: 2
```

We have the pass for CKV_AWS_135:

```
Check: CKV_AWS_135: "Ensure that EC2 is EBS optimized"
    PASSED for resource: tasks.amazon.aws.ec2_instance.Create the
temporary ec2 instance
        File: /roles/ec2tmp/tasks/main.yml:53-77
        Guide: https://docs.prismacloud.io/en/enterprise-edition/policy-
reference/aws-policies/aws-general-policies/ensure-that-ec2-is-ebs-
optimized
```

We also have the two false positives now showing:

```
Check: CKV_AWS_88: "EC2 instance should not have public IP."
    SKIPPED for resource: tasks.amazon.aws.ec2_instance.Create the
temporary ec2 instance
    Suppress comment: "While a public IP address is assigned to the
instance, it is locked down by the security group and the instance is
temporary."
        File: /roles/ec2tmp/tasks/main.yml:53-77
```

```
       Guide: https://docs.prismacloud.io/en/enterprise-edition/policy-
   reference/aws-policies/public-policies/public-12
```

For the second one, we have the following:

```
Check: CKV2_ANSIBLE_2: "Ensure that HTTPS url is used with get_url"
       SKIPPED for resource: tasks.ansible.builtin.get_url.download wp-
cli
       Suppress comment: "The URL passed in the variable is secured with
SSL/TLS protocol."
       File: /roles/wordpress/tasks/main.yml:11-18
```

As you can see, our comments are visible for all to see.

So, returning to the call-out at the start of the section, why have we covered this tool if Checkov doesn't have full coverage for Ansible? As you can see from the output of the scan of our playbook, while there is not much coverage now, each new release brings additional Ansible policies. Hence, as time goes on, coverage should only get more robust, and hopefully, we will bring this promising tool in line with the second tool we will look at: **KICS**, or to give it its full title, **keeping infrastructure as code secure**.

# Exploring KICS

KICS is another static code analysis tool, and like Checkov, it is open source. It is designed to help you find common misconfiguration issues, potential compliance issues, and even security vulnerabilities within your IaC code. It ships with support for Kubernetes, Docker, AWS CloudFormation, Terram, and, of course, Ansible, which we will be focusing on in this chapter.

KICS is designed to be easy to install, understand, and integrate into CI/CD pipelines. It includes over 2,400 customizable rules and is built for extensibility, allowing for the easy addition of support for new IaC tools and updates to existing integrations.

KICS is maintained and supported by **Checkmarx** specialists in software application security testing, meaning that KICS has a good pedigree.

## Running the scan

Let's dive straight in. If you haven't already, check out the example repository using the following command:

```
$ git clone https://github.com/russmckendrick/Learn-Ansible-Second-
Edition-Scan.git
```

Now, we can pull the latest container image from Docker Hub (https://hub.docker.com/r/checkmarx/kics) by using the following command:

```
$ docker image pull checkmarx/kics:latest
```

Change to the folder containing our Ansible playbook:

```
$ cd Learn-Ansible-Second-Edition-Scan
```

Then run the scan itself:

```
$ docker container run --rm --tty --volume ./:/ansible checkmarx/kics
scan --path /ansible/
```

As you can see, the docker command follows the same pattern we discussed when we ran Checkov up until where we pass the options to the KICS binary; here, we instruct KICS to run scan against the --path /ansible/, which is the directory we have mounting from our host machine inside the container using the --volume option.

## Reviewing the results

Now, let's take a look at the result of the scan; KICS presents its output, which is slightly different from Checkov in that the initial output is designed to give real-time information on the scan itself:

```
Scanning with Keeping Infrastructure as Code Secure v1.7.12
Preparing Scan Assets: Done
Executing queries: [------------------------------] 100.00%
Files scanned: 33
Parsed files: 32
Queries loaded: 292
Queries failed to execute: 0
```

Let's now work through the various results and group them by severity levels.

### Info and low-severity results

The first result highlights potentially risky file permissions for the files we create (using templates) or copy:

```
Risky File Permissions, Severity: INFO, Results: 5
Description: Some modules could end up creating new files on disk with
permissions that might be too open or unpredictable
Platform: Ansible
Learn more about this vulnerability: https://docs.kics.io/latest/
queries/ansible-queries/common/88841d5c-d22d-4b7e-a6a0-89ca50e44b9f
```

It then goes on to list all the affected files; here is a snippet of the first few:

```
[1]: ../../ansible/destroy.yml:52
     051:      - name: "Delete the group_vars/generated_aws_endpoints.
yml file"
```

```
052:           ansible.builtin.file:
053:               path: "group_vars/generated_aws_endpoints.yml"
```

Here is another:

```
[2]: ../../ansible/destroy.yml:81

   080:      - name: "Delete the group_vars/generated_efs_targets.yml
file"
   081:           ansible.builtin.file:
   082:               path: "group_vars/generated_efs_targets.yml"
```

Moving on to the next issue, we see the following:

```
Unpinned Package Version, Severity: LOW, Results: 1
Description: Setting state to latest performs an update and installs
additional packages possibly resulting in performance degradation or
loss of service
Platform: Ansible
Learn more about this vulnerability: https://docs.kics.io/latest/
queries/ansible-queries/common/c05e2c20-0a2c-4686-b1f8-
5f0a5612d4e8
```

Again, here is a sample of where it has spotted the issue:

```
[1]: ../../ansible/roles/stack_install/tasks/main.yml:8
   007:      name: "*"
   008:      state: "latest"
   009:      update_cache: true
```

The next and final low-scoring result is as follows:

```
EFS Without Tags, Severity: LOW, Results: 1
Description: Amazon Elastic Filesystem should have filesystem tags
associated
Platform: Ansible
Learn more about this vulnerability: https://docs.kics.io/latest/
queries/ansible-queries/aws/b8a9852c-9943-4973-b8d5-77dae9352851
```

Here are the details:

```
[1]: ../../ansible/destroy.yml:75
   074:      - name: "Delete the EFS File System"
   075:        community.aws.efs:
   076:          name: "{{ efs_name }}"
```

Let's quickly review the low-scoring ones before moving on to the one result with a medium score.

So, the first result was, **"Some modules could end up creating new files on disk with permissions that might be too open or unpredictable"**. It called out 11 places within our playbook where this could be an issue, so we should look at resolving these.

First off, if you ran the full scan, you will have noticed that three of the results are from the `destroy.yml` file.

Given that these tasks are removing files, we don't care about the file permissions here. So, rather than adding the permissions to the individual tasks, we should instruct KICS not to run the check across the whole file.

To do this, we need to add the following comment at the very top of the file:

```
# kics-scan disable=88841d5c-d22d-4b7e-a6a0-89ca50e44b9f
```

Next, we have `ansible.builtin.template` in `roles/efs/tasks/main.yml`. Rather than skip the test, I added the permissions using the `mode` key:

```
- name: "Generate the efs targets vars file"
  ansible.builtin.template:
    src: "targets.j2"
    dest: "group_vars/generated_efs_targets.yml"
    mode: "0644"
```

The final result is for the `ansible.builtin.get_url` module used by the task, which downloads `wp-cli` in the `roles/wordpress/tasks/main.yml` file.

When reviewing the code, it looked like the following:

```
- name: "Download wp-cli"
  ansible.builtin.get_url:
    url: "{{ wp_cli.download }}"
    dest: "{{ wp_cli.path }}"
```

This was immediately followed by the following:

```
- name: "Update permissions of wp-cli to allow anyone to execute it"
  ansible.builtin.file:
    path: "{{ wp_cli.path }}"
    mode: "0755"
```

Here, KICS highlights that we can set the mode as part of `ansible.builtin.get_url`, which means we do not have to do it separately, so I added the following to the download task:

```
    mode: "0755"
```

Then, I removed the second task. That clears the file permission issues KICS reported.

The next LOW score says, **"Setting state to latest performs an update and installs additional packages possibly resulting in performance degradation or loss of service"**.

This appears in `roles/stack-install/tasks/main.yml`, where the task uses `ansible.builtin.apt` to update the installed images, as this task is only called when we bootstrap our temporary EC2 instance and we made allowances for changes to the PHP version in the main playbook. I think it is safe to accept this as a false positive, so we can tell KICS not to run the test on this file by adding the following to the very top of `roles/stack-install/tasks/main.yml`:

```
# kics-scan disable=c05e2c20-0a2c-4686-b1f8-5f0a5612d4e8
```

This leaves us with **"Amazon Elastic Filesystem should have filesystem tags associated"**; the task it is complaining about is in `destroy.yml`, so the lack of tags does matter.

Let's exclude the check from being run. To do this, we need to append it to the end of the comment we already added, which means the comment at the end of `destroy.yml` now reads this:

```
# kics-scan disable=88841d5c-d22d-4b7e-a6a0-89ca50e44b9f,b8a9852c-
9943-4973-b8d5-77dae9352851
```

When appending IDs, please ensure that a comma separates them; otherwise, KICS will read them as one string. Finally, we have the high-severity results.

### High-severity results

Luckily, here we have just two problems called out across four tasks, starting with the following:

```
EFS Not Encrypted, Severity: HIGH, Results: 2
Description: Elastic File System (EFS) must be encrypted
Platform: Ansible
Learn more about this vulnerability: https://docs.kics.io/latest/
queries/ansible-queries/aws/727c4fd4-d604-4df6-a179-7713d3c85e20
```

These are the two tasks:

```
[1]: ../../ansible/roles/efs/tasks/main.yml:25
      024: - name: "Create the EFS File System"
      025:   community.aws.efs:
```

The second is in the `destroy.yml` file:

```
[2]: ../../ansible/destroy.yml:77
      076:     - name: "Delete the EFS File System"
      077:         community.aws.efs:
```

I think you can probably guess how we are going to resolve the second one; let's get it to ignore the test in `destroy.yml`:

```
# kics-scan disable=88841d5c-d22d-4b7e-a6a0-89ca50e44b9f,b8a9852c-
9943-4973-b8d5-77dae9352851,050f085f-a8db-4072-9010-
2cca235cc02f,727c4fd4-d604-4df6-a179-7713d3c85e20
```

For `roles/efs/tasks/main.yml`, the recommendation is to enable encryption, so let's take that advice:

```
- name: "Create the EFS File System"
  community.aws.efs:
    encrypt: true
    name: "{{ efs_name }}"
```

As you can see from the preceding snippet, we have added the `encrypt` parameter and set it to `true`.

The next issue highlighted by KICS also has to do with EFS filesystem encryption:

```
EFS Without KMS, Severity: HIGH, Results: 2
Description: Amazon Elastic Filesystem should have filesystem
encryption enabled using KMS CMK customer-managed keys instead of AWS
managed-keys
Platform: Ansible
Learn more about this vulnerability: https://docs.kics.io/latest/
queries/ansible-queries/aws/bd77554e-f138-40c5-91b2-2a09f878608e
```

The results are for the same files as the previous issue, so we will append the ID to the list of checks to disable at the top of the `destroy.yml` file.

Given that this is just a demo environment, I am happy to accept the potential risk of not using a customer-managed key vault to store my own managed encryption keys; so, in this instance, I will add the following:

```
# kics-scan disable=bd77554e-f138-40c5-91b2-2a09f878608e
```

I'll do so at the very top of the `roles/efs/tasks/main.yml` file. If this were a fixed production environment, then I would have added a role to launch and maintain AWS Key Management Service (`https://aws.amazon.com/kms/`) as part of the deployment.

### The results summary

The final part of the rules gives an overview of everything we have covered, which, for the initial scan with none of the fixes in place, is as follows:

```
Results Summary:
HIGH: 4
MEDIUM: 0
```

```
LOW: 2
INFO: 5
TOTAL: 11
```

## Re-running the scan

As before, there is a branch containing all of the updated files we discussed and implemented in the previous section; to change to it, run the following:

```
$ git switch chapter13-kics
```

You can then run the scan again using this:

```
$ docker container run --rm --tty --volume ./:/ansible checkmarx/kics
scan --path /ansible/
```

This should now return a clean bill of health:

```
Scanning with Keeping Infrastructure as Code Secure v1.7.13
Preparing Scan Assets: Done
Executing queries: [----------------------------] 100.00%
Results Summary:
HIGH: 0
MEDIUM: 0
LOW: 0
INFO: 0
TOTAL: 0
```

As you can see, no problems are being reported now.

## Output files

Before we finish the chapter, there is one more thing that we should quickly discuss about KICS: its ability to output a report in various file formats.

If you were to re-run the scans against the main and kics branches but using the following command, then you will notice that a file called results.html appears in your repo folder:

```
$ docker container run --rm --tty --volume ./:/ansible checkmarx/kics
scan --path /ansible/ --report-formats "html" --output-path /ansible/
```

As you can see, we are passing in two new flags; the first, --report-formats, tells KICS to output a report as an html file, and the second, --output-path, lets KICS know where to save the report file; in our case, as we are running KICS in a container that needs to be a location within the container that persists, once the container has finished running, the container will automatically be removed along with any files written.

When running the command against the main branch, which does not contain any of the fixes, we applied the header of the report, which looks like the following:

Figure 13.1 – Viewing the report showing issues

Then, re-running the scan against the KICS branch updates it to the following:

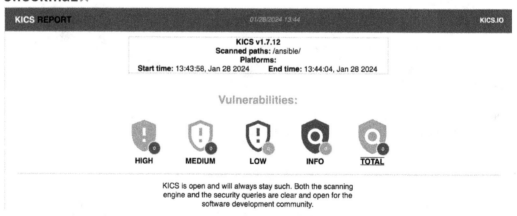

Figure 13.2 – A clean bill of health

You can also output in PDF, JSON, and other standard reporting formats. As you can see, this is a little more digestible than reading the output of the command line report we covered in the previous section.

We will put these reports to good use when we get to *Chapter 15, Using Ansible with GitHub Actions and Azure DevOps*, as we will publish the results as part of our pipeline runs.

# Summary

In this chapter, we have covered two tools that we can add to our workflows, and we manually ran scans against the playbook we developed in *Chapter 11, Highly Available Cloud Deployments*. As mentioned in the chapter, Checkov's support for Ansible is relatively new, so it has a different coverage to KICS. However, I am sure you agree that both tools worked well.

> **Important note**
>
> There is one elephant in the room, though; even without the same coverage level, both tools came up with slightly different results, so you should never rely on them 100% to fully secure your deployments. Think of them as trusted colleagues reviewing your code for anything obvious that stands out as being an issue rather than a security-focused cloud platform architect with a working knowledge of your workload who dictates precisely what measures you should take when deploying your infrastructure in a secure way fully.

As already mentioned at the end of the previous section, we will be revisiting KICS in *Chapter 15, Using Ansible with GitHub Actions and Azure DevOps*. Before we get there, now that we have looked at how we can review and secure our playbook code, we can now look at how we can secure our workload by quickly applying security best practices to the host operating systems that we are targeting using Ansible.

# Further reading

For more information about the tools and their maintainers, see the following links:

- **Checkov**: https://www.checkov.io/

- **Prisma Cloud**: https://www.paloaltonetworks.com/prisma/cloud/

- **KICS**: http://kics.io/

- **Checkmarx**: https://checkmarx.com/

# 14

# Hardening Your Servers Using Ansible

One of the advantages of using an orchestration and configuration tool such as Ansible is that it can be utilized to generate and deploy a complex set of configurations in a repeatable task across many hosts. In this chapter, we will look at a tool that scans your hosts using Ansible, dynamically generates a remediation playbook, and then runs it for you.

We will also look at running two different security tools that scan the WordPress installation we have used throughout the previous chapters.

This chapter covers the following topics:

- The scanning tools
- The playbook

## Technical requirements

Following our excursion into the cloud, we will return to our local machine and launch an Ubuntu 22.04 virtual machine using Multipass; as we will be running a workload that requires a bit more disk space than we have been used to so far, we will be altering the specs of the virtual machine when we launch it to increase the disk space and RAM.

As we will be installing a lot of different software on the virtual machine, your Multipass virtual machine will need to be able to download packages from the internet; there will be around 3 GB of various packages and configuration files to download.

You can find a complete copy of the playbook accompanying this chapter in the repository at `https://github.com/PacktPublishing/Learn-Ansible-Second-Edition/tree/main/Chapter14`.

# The scanning tools

Before we dive into the Playbook, let's quickly look at the three tools we will be running, starting with the one that does the most, **OpenSCAP**.

## OpenSCAP

First, we will be looking at one of Red Hat's tools, called OpenSCAP. Before we continue, the next section will contain many abbreviations.

So, what is SCAP? The **Security Content Automation Protocol** (**SCAP**) is an open standard that encompasses several components, all of which are open standards themselves, to build a framework that allows you to automatically assess and remediate your hosts against the **National Institute of Standards and Technology** (**NIST**) *Special Publication 800-53*.

This publication is a catalog of controls applied to all U.S. federal IT systems, apart from those maintained by the **National Security Agency** (**NSA**). These controls have been effected to help implement the **Federal Information Security Management Act** (**FISMA**) of 2002 across U.S. federal departments.

SCAP is made up of the following components:

- **Asset Identification** (**AID**) is a data model used for asset identification.

- **Asset Reporting Format** (**ARF**) is a vendor-neutral and technology-agnostic data model for transporting information on assets between different reporting applications and services.

- **Common Configuration Enumeration** (**CCE**) is a standard database of recommended configurations for common software. Each recommendation has a unique identifier. At the time of writing, the database hadn't been updated for over a decade.

- **Common Configuration Scoring System** (**CCSS**) is the continuation of CCE. It is used for generating a score for various software and hardware configurations across all types of deployments.

- **Common Platform Enumeration** (**CPE**) identifies hardware assets, operating systems, and software in an organization's infrastructure. Once identified, this data can then be used to search other databases to threat-assess the asset.

- **Common Weakness Enumeration** (**CWE**) is a common language for dealing with and discussing the causes of weaknesses in system architecture, design, and code that may lead to vulnerabilities.

- **Common Vulnerabilities and Exposures** (**CVE**) is a database of publicly acknowledged vulnerabilities. Most system administrators and IT professionals will have encountered the CVE database at some point. Each vulnerability receives a unique ID; for example, most people will know CVE-2014-0160, also known as **Heartbleed**. The Heartbleed vulnerability was a severe security flaw in OpenSSL (a cryptographic software library) that allowed attackers to steal sensitive information, such as passwords and private keys, from the memory of affected systems by exploiting a bug in the OpenSSL's implementation of the **transport layer security** (**TLS**)/**datagram transport layer security** (**DTLS**) heartbeat extension.

- **Common Vulnerability Scoring System** (**CVSS**) is a method that helps capture the characteristics of a vulnerability to produce a normalized numerical score, which can then be used to describe the impact of a vulnerability, for example, low, medium, high, and critical.

- **Extensible Configuration Checklist Description Format** (XCCDF) is an XML format for describing security checklists. It can also be used for configuration and benchmarks and provides a common language for all the parts of SCAP.

- **Open Checklist Interactive Language** (**OCIL**) is a framework for expressing questions to an end user and the procedures to process the responses in a standardized way.

- **Open Vulnerability and Assessment Language** (**OVAL**) is defined in XML and aims to standardize the transfer of security content across all of the tools and services offered by NIST, the MITRE Corporation, the **United States Computer Emergency Readiness Team** (**US-CERT**), and the United States **Department of Homeland Security** (**DHS**).

- **Trust Model for Security Automation Data** (**TMSAD**) is an XML document that aims to define a common trust model that can be applied to the data being exchanged by all components that make up SCAP.

As you can imagine, thousands of man-years have gone into producing SCAP and its components to make its foundation. Some of the projects have been around in one form or another since the mid-90s, so they are well-established and considered the de facto standard when it comes to security best practices; however, I am sure you think that it all sounds very complicated – after all, these are standards that have been defined and are being maintained by scholars, security professionals, and government departments.

This is where OpenSCAP comes in. The OpenSCAP project, maintained by Red Hat and certified by NIST for supporting the SCAP standard, allows you to apply all the best practices we have discussed using a command-line client.

> **Note**
>
> The automatic remediation scripts in OpenSCAP are a work in progress, and there are known issues that we will address toward the end of the chapter. Because of this, your output may differ from that covered in this chapter.

OpenSCAP, like many Red Hat projects, has support for Ansible, and the current release introduces support for automatically generating Ansible playbooks to remediate non-conformance discovered during an OpenSCAP scan.

The next two tools we will be looking at will be scanning our WordPress site, starting with WPScan.

## WPScan

The second tool we will be running is called **WPScan** and we will use it to scan our WordPress site. WPScan is a command-line tool that can perform various security assessments and vulnerability tests on WordPress installations. It can detect common configuration errors, outdated themes, weak passwords, and other potential risks. WPScan is easy to install – especially as we will be using the container version and running it using Docker, which we will also be going for the third and final tool, OWASP ZAP.

## OWASP ZAP

Web vulnerabilities such as SQL injection, cross-site scripting, broken authentication, and insecure deserialization can threaten our WordPress site's security and quality. To help identify and prioritize such vulnerabilities, we can use **OWASP ZAP**. This tool, the third and final one we will cover in the chapter, generates reports, alerts, and graphs that assist us in visualizing and addressing the findings. Moreover, OWASP ZAP is user-friendly and easy to install, making it a valuable resource for enhancing our site's security and overall quality.

# The playbook

We will split the playbook into a few different roles to run the various scanning tools that will be running in the chapter – as you can see from the `site.yml` file, we are adding some conditions to the roles containing our tasks. The start of the file looks like all of the other playbook files we have been running:

```
- name: "Scan our WordPress Ansible Playbook and stack"
  hosts: ansible_hosts
  gather_facts: true
  become: true
  become_method: "ansible.builtin.sudo"
  vars_files:
    - 'group_vars/common.yml'
```

As mentioned, roles are where this playbook starts to differ from the previous playbooks we have been running up to this point in the book.

As you can see from the following source, we are defining tags alongside the roles themselves:

```
  roles:
    - { role: 'common', tags: ['openscap','scan'] }
    - { role: 'docker', tags: ['docker','scan'] }
```

As you can see, we are using the openscap, scan, and docker tags followed by wordpress, which used the roles directly from *Chapter 5, Deploying WordPress*:

```
- { role: 'stack_install', tags: ['wordpress'] }
- { role: 'stack_config', tags: ['wordpress'] }
- { role: 'wordpress', tags: ['wordpress'] }
```

Finally, we have roles that run scans and openscap:

```
- { role: 'scan', tags: ['scan'] }
- { role: 'openscap', tags: ['openscap'] }
```

So, what does this mean? Well, later in the chapter, when it comes to running the playbook, we will only be running specific roles; for example, to run OpenSCAP, we will use the following commands:

```
$ ansible-playbook -i hosts site.yml --tags "openscap" --extra-vars
"scap_options_remediation=true"
$ ansible-playbook -i hosts site.yml --tags "openscap"
```

When running the first command, it will run just the common and openscap roles and run the remediation Ansible Playbook and bash script, both of which will be automatically generated during the initial scan – it will also download a copy of the results, an implementation guide, a copy of the playbook, and a copy of the bash scripts.

The second of the two commands will rerun the scan host and download a copy of the results again.

Once we have finished running OpenSCAP, we will then redeploy our host and run the following:

```
$ ansible-playbook -i hosts site.yml --tags "wordpress"
```

This, as I am sure you will have guessed, will run the three wordpress roles. Then, with WordPress installed, we can run the following:

```
$ ansible-playbook -i hosts site.yml --tags "scan"
```

This will execute the common, docker, and scan roles.

We can also run these commands to run just one of the two scanning tools that the scan role runs:

```
$ ansible-playbook -i hosts site.yml --tags "scan" --extra-vars "scan_
types=zap"
$ ansible-playbook -i hosts site.yml --tags "scan" --extra-vars "scan_
types=wpscan"
```

But we are getting ahead of ourselves; let's work our way through the preceding roles before we think about running the playbook.

## The common role

This role contains a single task in `roles/common/tasks/main.yml`, and its only job is to set a fact containing the current date and time:

```
- name: "Set a fact for the date"
  ansible.builtin.set_fact:
    the_date: "{{ lookup('pipe', 'date +%Y-%m-%d-%H%M') }}"
```

You might think, *"That seems a little basic."* However, as we will be using the `the_date` variable several times throughout the roles in this playbook, we only want it to be generated once as it will be used to create file and folder names that are then called later in tasks.

If we use `{{ lookup('pipe', 'date +%Y-%m-%d-%H%M') }}` to insert the date dynamically as part of other variables and tasks, we need to be cautious. This is because some parts of the playbook can take several minutes to finish running.

For instance, we may create a file called `myfile-2024-02-16-1300.yml` at one point in the playbook. However, if we dynamically set the date and time, and several tasks later, it takes five minutes for the playbook to get to that task, we could reference a file called `myfile-2024-02-16-1305.yml`. This would result in an error as the file does not exist. Therefore, we should only use the date and time lookup once during the playbook run.

## The Docker role

This role contains all of the tasks and variables needed to install and configure Docker on our target host, much like the roles discussed in *Chapter 4, Deploying a LAMP Stack*, and *Chapter 5, Deploying WordPress*; this role uses the `ansible.builtin.apt`, `ansible.builtin.apt_key`, and `ansible.builtin.apt_repository` modules to do the following:

1.  Download and install the prerequisites required for Docker to run.
2.  Add the **GNU Privacy Guard** (**GPG**) key for the official Docker **advanced packaging tool** (**APT**) repository.
3.  Configure the official Docker APT repository.
4.  Install Docker itself along with the Docker command-line tool.
5.  Ensure that Docker is running and set to start on boot.

To review the full list of tasks and variables for this role, see the following:

*   `https://github.com/PacktPublishing/Learn-Ansible-Second-Edition/blob/main/Chapter14/roles/docker/defaults/main.yml`
*   `https://github.com/PacktPublishing/Learn-Ansible-Second-Edition/blob/main/Chapter14/roles/docker/tasks/main.yml`

Next, we have the roles that install WordPress.

## The WordPress roles

As you have already seen from the `site.yml` file at the start of the Playbook section of this chapter, here, we are just reusing the roles that we discussed at length in *Chapter 5*, *Deploying WordPress*. If you want to review these, you can see them at `https://github.com/PacktPublishing/Learn-Ansible-Second-Edition/tree/main/Chapter05/roles`.

## The scan role

As already mentioned, we will be using Docker to run WPScan and OWASP ZAP; this allows us to reuse the same tasks. Let's look at `roles/scan/tasks/main.yml`; first, we need to pull the Docker image or images:

```
- name: "Pull the Docker image for the scanning tool"
  community.docker.docker_image:
    name: "{{ item.image }}"
    source: "{{ item.source }}"
  loop: "{{ scan }}"
  when: "item.name in scan_types"
  loop_control:
    label: "{{ item.name }}"
```

We are switching it up slightly in that we are using `loop` rather than `with_items`; this gives more control over what happens when looping through. In this task, we are using `label` to show which of the scanning tools is currently processing.

You may also notice that we have a `when` condition; this allows us to run both of the scans or just one of the two by passing in the name of the scan in the `scan_types` variable. When we look at the variables in a moment, you will see that by default, we are passing in the names of both scanning tools.

This pattern of `loop`, `loop_control`, and `when` will be repeated throughout all the tasks in this role. We have a task that will create a folder on the virtual machine; we will be mounting this folder into the container at runtime so that we can keep a copy of the scan output:

```
- name: "Create the folder which we will mount inside the container"
  ansible.builtin.file:
    path: "{{ item.log.remote_folder }}"
    state: "directory"
    mode: "0777"
  loop: "{{ scan }}"
  when: "item.name in scan_types"
  loop_control:
    label: "{{ item.name }}"
```

Now, with the container image and folder created, we can run the scan:

```
- name: "Run the scan"
  community.docker.docker_container:
    detach: "{{ item.detach }}"
    auto_remove: "{{ item.auto_remove }}"
    name: "{{ item.name }}"
    volumes: "{{ item.log.remote_folder }}:{{ item.container_folder
}}"
    image: "{{ item.image }}"
    command: "{{ item.command }}"
  register: docker_scan
  ignore_errors: true
  no_log: true
  loop: "{{ scan }}"
  when: "item.name in scan_types"
  loop_control:
    label: "{{ item.name }}"
```

As you can see, everything is being passed to the container as variables; this is how we can run two very different tools with a single common task, and more so later when we look at the variables.

You will have also noted that we are adding a few options to the end of this task; these are as follows:

- `register`: Here, we are just registering the output of the task – nothing special here

- `ignore_errors`: This tells Ansible to continue running should it detect an error; in our case, the containers we are running will purposely trigger an error code as they have been designed to halt and not proceed with any further tasks until the scan does not fail

- `no_log`: This suppresses the output – as we save the output when running the scan, we do not need the output printed to the terminal when we run the task

As we are registering an output, the next task is a debug line. This follows the same pattern as debug tasks in other chapters, so we will be moving to the task that downloads a copy of the reports:

```
- name: "Download the report"
  ansible.builtin.fetch:
    src: "{{ item.log.remote_folder }}{{ item.log.file }}"
    dest: "{{ item.log.local_folder }}"
    flat: true
    mode: "0644"
  loop: "{{ scan }}"
  when: "item.name in scan_types"
  loop_control:
    label: "{{ item.name }}"
```

This uses the `ansible.builtin.fetch` module setting the `flat` option to `true`. This option copies the file rather than the full directory path. The final task removes the container, meaning that when we next run a scan, it will start from scratch and spawn a new container rather than reusing the one we have just finished using:

```
- name: "Remove the scan container"
  community.docker.docker_container:
    name: "{{ item.name }}"
    state: "absent"
  loop: "{{ scan }}"
  when: "item.name in scan_types"
  loop_control:
    label: "{{ item.name }}"
```

Now that we know what the tasks look like, let us look at the variables, which can be found in `roles/scan/defaults/main.yml`. The first variable sets the scan we want to run, and as already mentioned, this gives the name of the two scans:

```
scan_types:
  - "{{ common_scan_settings.dict.wpscan }}"
  - "{{ common_scan_settings.dict.zap }}"
```

Next up in `roles/scan/defaults/main.yml`, we have a block of variables that could be commonly used across both scanning tools:

```
common_scan_settings:
  detach: false
  auto_remove: false
  source: "pull"
  local_folder: "output/"
  report_name: "{{ the_date }}-results-"
  dict:
    wpscan: "wpscan"
    zap: "zap"
```

Finally, we have the primary `scan` variable, which is the one we have been looping over; it starts with WPScan:

```
scan:
  - name: "{{ common_scan_settings.dict.wpscan }}"
    image: "wpscanteam/wpscan:latest"
    source: "{{ common_scan_settings.source }}"
    detach: "{{ common_scan_settings.detach }}"
    auto_remove: "{{ common_scan_settings.auto_remove }}"
    container_folder: "/tmp/{{ common_scan_settings.dict.wpscan }}/"
```

```
    command: "--url http://{{ ansible_host }} --enumerate u --plugins-
detection mixed --format cli-no-color --output /tmp/{{ common_scan_
settings.dict.wpscan }}/{{ common_scan_settings.report_name }}{{
common_scan_settings.dict.wpscan }}.txt"
    log:
      remote_folder: "/tmp/{{ common_scan_settings.dict.wpscan }}/"
      local_folder: "{{ common_scan_settings.local_folder }}"
      file: "{{ common_scan_settings.report_name }}{{ common_scan_
settings.dict.wpscan }}.txt"
```

The block that follows is the one for OSWAP ZAP:

```
  - name: "{{ common_scan_settings.dict.zap}}"
    image: "ghcr.io/zaproxy/zaproxy:stable"
    source: "{{ common_scan_settings.source }}"
    detach: "{{ common_scan_settings.detach }}"
    auto_remove: "{{ common_scan_settings.auto_remove }}"
    container_folder: "/zap/wrk/"
    command: "zap-baseline.py -t http://{{ ansible_host }} -g gen.conf
-r {{ common_scan_settings.report_name }}{{ common_scan_settings.dict.
zap }}.html"
    log:
      remote_folder: "/tmp/{{ common_scan_settings.dict.zap }}/"
      local_folder: "{{ common_scan_settings.local_folder }}"
      file: "{{ common_scan_settings.report_name }}{{ common_scan_
settings.dict.zap }}.html"
```

As you can see, we pass in the different container images and commands to run the scan while using the same variables. Because of this, we could keep the tasks used in the role completely neutral, meaning that we didn't have to consider anything custom to the tool we were running.

That concludes the scan role, leaving us with, as I am sure you will have already guessed from how long the tool explanation was at the start of the chapter, the most complex role in the playbook: OpenSCAP.

## The OpenSCAP role

When writing a playbook, it is essential to know how the tool you are automating works; given that OpenSCAP is a little complex, let's review the steps needed to manually run a scan and remediate the problems it finds using an automatically generated Ansible playbook and a shell script.

> **Note**
>
> While the commands to run OpenSCAP follow, you do not need to follow along; these are provided to illustrate the process we need to follow in our Playbook role.

First, we need to download and install OpenSCAP itself, along with a few tools we will also need:

```
$ sudo apt-get install unzip curl libopenscap8
```

Next up, we need to download the actual content – these definitions cover several different operating systems and various levels of compliance. The GitHub repository for this content can be found at `https://github.com/ComplianceAsCode/content`, and at the time of writing, the current release is 0.1.71.

Get the release URL for the zip file, which contains the files we need from the releases page, then download and unzip on the host:

```
$ wget https://github.com/ComplianceAsCode/content/releases/download/
v0.1.71/scap-security-guide-0.1.71.zip
unzip scap-security-guide-0.1.71.zip
```

Now that we have OpenSCAP and the definition files installed, we can get some information on what is available for our Ubuntu 22.04 operating system:

```
$ sudo oscap info --fetch-remote-resources scap-security-guide-0.1.71/
ssg-ubuntu2204-ds.xml
```

This will give us the name of the profile we want to use; in our case, it is `xccdf_org.ssgproject.content_profile_cis_level1_server`. Once we have this, we can run the scan itself:

```
$ oscap xccdf eval --profile xccdf_org.ssgproject.content_profile_cis_
level1_server  --results-arf result.xml --report report.html scap-
security-guide-0.1.71/ssg-ubuntu2204-ds.xml
```

This will generate two output files: an HTML copy of a report containing everything that needs fixing in a nicely digestible format we can read, and a second XML file containing the same information in a format OpenSCAP can read.

We can then take the XML file and generate a more detailed guide on how we could resolve the issues found by running the following:

```
$ sudo oscap xccdf generate guide  --profile xccdf_org.ssgproject.
content_profile_cis_level1_server scap-security-guide-0.1.71/
ssg-ubuntu2204-ds.xml  > guide.html
```

However, as this book is about Ansible, it would be better to have a Playbook to fix as many of the issues as possible, and running the following command will give us just that:

```
$ sudo oscap xccdf generate fix --fetch-remote-resources --fix-type
ansible --result-id "" result.xml > playbook.yml
```

Finally, not everything can be resolved using the Playbook method, so having a bash script to fix any issues that can't be resolved by running the playbook is also a great idea as it will mean less manual work for us to do:

```
$ sudo oscap xccdf generate fix --fetch-remote-resources --fix-type
bash --result-id "" result.xml > bash.sh
```

Now we have the Playbook and bash script; we need to run them, copy the playbook to our local machine, and run it using the following:

```
$ ansible-playbook -i hosts --become -become-method=sudo output/
ansiblevm-playbook.yml
```

Then we go back to the virtual machine, and run the bash script using the following:

```
$ sudo bash bash.sh
```

You will have seen a lot of output, but if everything goes as planned when you rerun the scan, you should see a lot of issues being reported.

> **Note**
>
> The code in the repo contains the variables and tasks for a feature we will not cover here, as the content we are downloading from GitHub can take up a lot of space on your drive. These tasks are included to remove any unneeded files.

So, now that we have an idea of the steps we need to automate, let's dive straight in.

First, let's look at the variables, which can be found in `roles/openscap/default/main.yml`, and that we will be using within our tasks.

Start with the option that, if set to `true`, will execute the remediation Playbook and Bash script:

```
scap_options_remediation: false
```

Next, we have the packages needed to run OpenSCAP and OpenSCAP itself:

```
scap_packages:
  - "unzip"
  - "curl"
  - "libopenscap8"
```

Then we have information to download the content from GitHub; note that we are passing the API URL and not the direct download link (more on why later in the chapter):

```
openscap_download:
  openscap_github_release_api_url: "https://api.github.com/repos/
ComplianceAsCode/content/releases/latest"
  dest: "/tmp/scap-security-guide"
```

Now we have a long list of filenames and details on the profile we need to use:

```
openscap_scan:
  ssg_file_name: "{{openscap_download.dest}}/ssg-{{ ansible_facts.
distribution | lower }}{{ ansible_facts.distribution_version |
replace('.','') }}-ds.xml"
  profile_search: "cis_level1_server"
  output_dir: "/tmp/"
  output_file_xml: "{{ inventory_hostname }}-result.xml"
  output_file_html: "{{ inventory_hostname }}-report.html"
  output_file_guide: "{{ inventory_hostname }}-guide.html"
  output_file_playbook: "{{ inventory_hostname }}-playbook.yml"
  output_file_bash: "{{ inventory_hostname }}-bash.sh"
  local_output_dir: "output/{{ the_date }}-openscap-results"
```

Notice that we are trying not to hardcode any values; for example, when referring to the operating system, we use `{{ ansible_facts.distribution | lower }}{{ ansible_facts.distribution_version | replace('.','') }}`, which, in our case, gives us ubuntu2204. This means that if OpenSCAP supports it, we can run our Playbook on other Ubuntu distributions without making any changes.

The tasks that use these variables can be found in `roles/openscap/tasks/main.yml`; we begin with two tasks that install OpenSCAP, the first of which makes sure that the APT cache and our operating system are both up to date:

```
- name: "Update apt cache and upgrade packages"
  ansible.builtin.apt:
    name: "*"
    state: "latest"
    update_cache: "yes"
```

The tasks immediately after installing OpenSCAP itself and the other packages we need:

```
- name: "Install common packages"
  ansible.builtin.apt:
    state: "present"
    pkg: "{{ scap_packages }}"
```

Now, we create the directory where we will be storing the OpenSCAP content we will be downloading from GitHub:

```
- name: "Create the directory to store the scap security guide
content"
  ansible.builtin.file:
    path: "{{ openscap_download.dest }}"
    state: "directory"
    mode: "0755"
```

With our destination folder in place, we can now download the content and unarchive it:

```
- name: "Download the latest scap security guide content"
  ansible.builtin.unarchive:
    src: "{{ lookup('url', '{{ openscap_download.openscap_github_
release_api_url }}', split_lines=false) | from_json | json_
query('assets[?content_type==`application/zip`].browser_download_url')
| last }}"
    dest: "{{ openscap_download.dest }}"
    creates: "{{ openscap_download.dest }}/README.md"
    list_files: true
    remote_src: true
  register: scap_download_result
```

On the face of it, while it looks a little complicated, there is quite a bit going on; let's break down how we are getting the value to populate into the `src` key.

We use Ansible's `lookup` plugin to fetch and process data from the GitHub API, giving us the latest release information for the OpenSCAP Content GitHub repository:

- `{{ lookup('url', '{{ openscap_download.openscap_github_release_api_url }}', split_lines=false) }}`: The `lookup` plugin is being used here with the `url` lookup type, which fetches data from the given URL that is specified by the `openscap_download.openscap_github_release_api_url` variable, which points to the API endpoint for the latest release of a GitHub repository (`https://api.github.com/repos/ComplianceAsCode/content/releases/latest`). The `split_lines=false` parameter ensures that the fetched content is not split into lines, preserving its JSON structure.

- `| from_json`: This part of the code takes the output from the `lookup` plugin, which is expected to be a JSON string, and converts it into an Ansible data structure (such as a dictionary or a list) that can be further processed.

- | json_query('assets[?content_type==`application/zip`].browser_download_url'): This uses the json_query filter with a JMESPath expression to query the converted JSON data. The 'assets[?content_type==`application/zip`].browser_download_url' query looks for items in the assets array where content_type is application/zip, and then extracts browser_download_url. This URL is typically used to directly download the asset from a browser.

- | last: Finally, the last filter is used to get the last URL from the list of URLs returned by the json_query filter. We are doing this as there might be multiple assets with the application/zip content type, but we are only interested in the most recent or last one listed.

This means that we do not have to hardcode the version number of the latest release into our Playbook, which is helpful as the OpenSCAP content repo is updated at least once every few weeks.

The other options we are passing to the ansible.builtin.unarchive module are as follows:

- dest: The destination directory on the target machine where the archive will be extracted is specified

- creates: This parameter is used as a conditional check to prevent re-downloading and extracting the archive if a particular file exists

- list_files: When set to true, this option lists all the files in the archive file; we will use this list to copy the files to our destination folder

- remote_src: Setting this to true indicates that the source archive is located on a remote server, not on the control machine running Ansible; this is needed to download content directly from a URL

The following two tasks move the files to the root of openscap_download.dest as they would have been unarchived to a folder containing the version number – which we don't want to use, as it could change between runs:

```
- name: "Move scap security guide content to the correct location"
  ansible.builtin.shell: "mv {{ openscap_download.dest }}/{{ scap_
download_result.files[0] }}/* {{ openscap_download.dest }}"
  when: scap_download_result.changed
- name: "Remove the downloaded scap security guide content"
  ansible.builtin.file:
    path: "{{ openscap_download.dest }}/{{ scap_download_result.
files[0] }}"
    state: "absent"
  when: scap_download_result.changed
```

Note that we are only running these tasks when the task that downloads the files has changed.

The final bit of information we need before we can run the OpenSCAP scan is which profile to use. To get this, we need to run the command to print information on the profiles available for our operating system:

```
- name: "Get information of the SCAP profiles available for the target
system"
  ansible.builtin.command: "oscap info –profiles –fetch-remote-
resources {{ openscap_scan.ssg_file_name }}"
  register: scap_info
```

Now that we have the information on the available profiles registered as `scap_info`, we can filter this list based on the contents of `openscap_scan.profile_search` and set a fact:

```
- name: "Extract profile name based on our selection criteria"
  ansible.builtin.set_fact:
    profile_name: "{{ scap_info.stdout_lines | select('search',
openscap_scan.profile_search) | map('regex_replace', '^(.*?):.*$',
'\\1') | first }}"
```

With the fact set, we can run the scan itself:

```
- name: "Run OpenSCAP scan"
  ansible.builtin.command: "oscap xccdf eval --profile {{ profile_
name }} --results-arf {{ openscap_scan.output_dir }}{{ openscap_scan.
output_file_xml }} --report {{ openscap_scan.output_dir }}{{ openscap_
scan.output_file_html }} {{ openscap_scan.ssg_file_name }}"
  ignore_errors: true
  no_log: true
  register: scap_scan
```

As you can see, we are suppressing the output by using `no_log: true`; this is because we don't really need to see the output at this stage and can ignore errors, like in the previous role where we ran WPScan and OSWAP ZAP.

Now that we have the output of the scan, we need to create a folder on our Ansible host to copy the output files to the following:

```
- name: "Ensure the local output directory exists"
  ansible.builtin.file:
    path: "{{ openscap_scan.local_output_dir }}"
    state: directory
    mode: "0755"
  delegate_to: "localhost"
  become: false
```

As you can see, we are using `delegate_to` to ensure that Ansible runs the task on `localhost`, and we are telling it not to become a privileged user.

Now we can `fetch` the `output.xml` and `report.html` files:

```
- name: "Copy the SCAP report and results file to local machine"
  ansible.builtin.fetch:
    src: "{{ item }}"
    dest: "{{ openscap_scan.local_output_dir }}/"
    flat: true
    mode: "0644"
  with_items:
    - "{{ openscap_scan.output_dir }}{{ openscap_scan.output_file_xml
}}"
    - "{{ openscap_scan.output_dir }}{{ openscap_scan.output_file_html
}}"
```

Next, we need to generate the guide and remediation files:

```
- name: "generate SCAP guide"
  ansible.builtin.command: "oscap xccdf generate guide --profile {{
profile_name }} {{ openscap_scan.ssg_file_name }}"
  ignore_errors: true
  register: scap_guide
```

You may have noticed we are not saving a file here; we are just registering the output. That is because all of the content for the guide is output to the screen when the command is run, so rather than direct the output to a file on the virtual machine and copy it, we can capture the output and then create a file on our local machine that contains this content, essentially a fancy *copy + paste* from the remote host to our local one:

```
- name: "Copy SCAP guide to local machine"
  ansible.builtin.copy:
    content: "{{ scap_guide.stdout }}"
    dest: "{{ openscap_scan.local_output_dir }}/{{ openscap_scan.
output_file_guide }}"
    mode: "0644"
  when: scap_guide is defined
  delegate_to: "localhost"
  become: false
```

This is then repeated for the remediation Ansible Playbook:

```
- name: "Generate SCAP fix playbook"
  ansible.builtin.command: "oscap xccdf generate fix --fetch-remote-
resources --fix-type ansible --result-id '' {{ openscap_scan.output_
dir }}{{ openscap_scan.output_file_xml }}"
  ignore_errors: true
  register: scap_playbook
```

```
- name: "Copy SCAP playbook to local machine"
  ansible.builtin.copy:
    content: "{{ scap_playbook.stdout }}"
    dest: "{{ openscap_scan.local_output_dir }}/{{ openscap_scan.
output_file_playbook }}"
    mode: "0644"
  when: scap_playbook is defined
  delegate_to: "localhost"
  become: false
```

Then again, for the remediation Bash script:

```
- name: "Generate SCAP fix bash script"
  ansible.builtin.command: "oscap xccdf generate fix --fetch-remote-
resources --fix-type bash --result-id '' {{ openscap_scan.output_dir
}}{{ openscap_scan.output_file_xml }}"
  ignore_errors: true
  register: scap_bash_script

- name: "Copy SCAP bash script to local machine"
  ansible.builtin.copy:
    content: "{{ scap_bash_script.stdout }}"
    dest: "{{ openscap_scan.local_output_dir }}/{{ openscap_scan.
output_file_bash }}"
    mode: "0644"
  when: scap_bash_script is defined
  delegate_to: "localhost"
  become: false
```

The remaining tasks in the role deal with the remediation work, starting with the playbook:

```
- name: "Run the remediation playbook"
  ansible.builtin.command: "ansible-playbook -i {{ inventory_file }}
--become --become-method sudo {{ openscap_scan.local_output_dir }}/{{
openscap_scan.output_file_playbook }}"
  when: scap_options_remediation
  delegate_to: "localhost"
  become: false
  register: remediation_playbook
```

Then, as we never kept a copy of the bash script on the target virtual machine, we need to copy it back there:

```
- name: "Copy the remediation bash script to the target machine"
  ansible.builtin.copy:
    src: "{{ openscap_scan.local_output_dir }}/{{ openscap_scan.
```

```
output_file_bash }}"
    dest: "{{ openscap_scan.output_dir }}"
    mode: "0755"
  when: scap_options_remediation
```

Once copied, we can run the script:

```
- name: "Run the remediation bash script"
  ansible.builtin.command: "bash {{ openscap_scan.output_dir }}{{
openscap_scan.output_file_bash }}"
  when: scap_options_remediation
  register: remediation_bash_script
```

With that task, the role is complete, and we now have all the pieces in place to run our playbook.

## Running the playbook

In *Chapter 1, Installing and Running Ansible*, we covered the installation and usage of Multipass; since then, we have been launching our local virtual machines using the same commands. In this chapter, as we need a little more disk space and RAM, we are going to be adding a few extra options when we launch the virtual machine:

```
$ multipass launch -n ansiblevm --cloud-init cloud-init.yaml --disk
10G --memory 4G
```

Once the virtual machine has launched, you can get the IP address of the host by running the following:

```
$ multipass info ansiblevm
```

Once you have the IP address, create a copy of hosts.example, calling its hosts and updating the IP address as we have done in previous chapters. Once your hosts inventory file is in place, we can start to run the playbook, starting with the OpenSCAP scan:

```
$ ansible-playbook -i hosts site.yml --tags "openscap" --extra-vars
"scap_options_remediation=true"
```

As you can see, we are running using the openscap tag and setting the scap_options_ remediation variable to true; if you recall, the default for this variable is false, meaning the remediation tasks will be executed during this playbook run.

Once completed, you will find several files in the output folder on your local machine; if you are not following along, then you can find a copy of the output at https://github.com/ PacktPublishing/Learn-Ansible-Second-Edition/tree/main/Chapter14/ examples/01-scap_options_remediation_true.

As you can see from the following screen, on the initial run, we had 98 failed results:

Figure 14.1 – The initial results

As we ran the remediation tasks as part of the playbook run, we know that the score should now be improved, so let's rerun the playbook – this time skipping the remediation tasks altogether:

```
$ ansible-playbook -i hosts site.yml --tags "openscap"
```

Once completed, you should have another folder of results; again, you can view the results at https://github.com/PacktPublishing/Learn-Ansible-Second-Edition/tree/main/Chapter14/examples/02-scap_options_remediation_false:

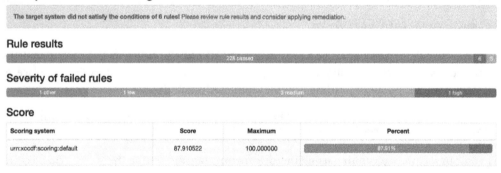

Figure 14.2 – The updated results

As you can see, this has dramatically improved the score, and we only have six failures this time.

Next, we need to install WordPress; let's start afresh with that. To make a fresh start, run the following commands to terminate the virtual machine and replace it with a new one:

```
$ multipass stop ansiblevm
$ multipass delete --purge ansiblevm
```

```
$ multipass launch -n ansiblevm --cloud-init cloud-init.yaml --disk
10G --memory 4G
$ multipass info ansiblevm
```

Update the hosts file with the new IP address and then run the following command to install WordPress:

```
$ ansible-playbook -i hosts site.yml --tags "wordpress"
```

With WordPress installed, you can run the WPScan and OSWAP ZAP scans with the following command:

```
$ ansible-playbook -i hosts site.yml --tags "scan"
```

Once completed, you will have the scan results in the output folder; you can find examples of the results at https://github.com/PacktPublishing/Learn-Ansible-Second-Edition/ tree/main/Chapter14/examples. The folder also contains the entire output from each of the playbook runs so far up to this point in the chapter.

Also, as mentioned at the start of the chapter, you can run each of the scans independently of each other using the following commands:

```
$ ansible-playbook -i hosts site.yml --tags "scan" --extra-vars "scan_
types=zap"
$ ansible-playbook -i hosts site.yml --tags "scan" --extra-vars "scan_
types=wpscan"
```

Once you have finished running the playbooks, you can remove the virtual machine by running the following:

```
$ multipass stop ansiblevm
$ multipass delete --purge ansiblevm
```

With the virtual machine cleaned up, that concludes our look at using Ansible to scan and harden our server.

Before we move on to the next chapter, I recommend you look at the remediation playbook, which was generated when we first ran OpenSCAP.

It can be found at https://github.com/PacktPublishing/Learn-Ansible-Second-Edition/blob/main/Chapter14/examples/01-scap_options_remediation_true/ ansiblevm-playbook.yml, and as you can see, it contains over 4,600 lines of code!

# Summary

In this chapter, we generated a playbook to remediate any CIS level-1 non-compliance errors found during a scan. As well as being cool, it is also convenient if you imagine you are running a few dozen servers that all need to be compliant and that all need an entire audit history.

You now have the foundations of a playbook that you can use to target those hosts daily, audit them, and store the results away from the host itself. Also, if you need to, depending on your configuration, you have a way of automatically resolving any non-conformance found during the scan.

We also ran scans against our WordPress installation and again stored the results away from the host itself – while the WPScan and OSWAP ZAP scans didn't include any remediation, you could quickly review the results and update your WordPress deployment script to remediate the issues raised at deployment time.

So far, we have been running our Ansible Playbooks from our local machine; in the next chapter, it is time to move from running our Ansible code from our local machines into the cloud and look at how we can use Azure DevOps Pipelines and GitHub Actions to execute our playbooks.

# 15

# Using Ansible with GitHub Actions and Azure DevOps

In this chapter, we will start running Ansible in the cloud instead of our local machines, something we have been doing up to this point.

First, this chapter will examine two services I often use during my day job:

- Running GitHub Actions
- Running pipelines in Azure DevOps

Before we move on, we will examine tools designed to execute Ansible from a central location in *Chapter 16, Introducing Ansible AWX and Red Hat Ansible Automation Platform.*

Neither of the two services we will be looking at has what you would call native support for Ansible; however, they both provide ephemeral compute resources that can be configured using YAML, which you can ship alongside your playbook code.

This chapter will cover a more complex playbook in both **GitHub Actions** and **Azure DevOps**. We will also discuss some considerations when running Ansible away from your machine.

So, rather than discussing it anymore, let's dive straight in and look at GitHub Actions.

## Technical requirements

If you are following along with the example code that we will be working through, then you will need access to a GitHub and an Azure DevOps account, as well as an Azure account since we will be launching a WordPress instance running in Azure as part of this chapter.

You can find a complete copy of the playbook, GitHub Action config, and Azure DevOps Pipeline code accompanying this chapter in this book's GitHub repository at `https://github.com/PacktPublishing/Learn-Ansible-Second-Edition/tree/main/Chapter15`.

# GitHub Actions

GitHub Actions is a comprehensive platform for **continuous integration (CI)** and **continuous delivery (CD)** from GitHub. It enables you to automate your build, test, and deployment pipeline while hosting your code and GitHub's exhaustive suite of code management tools. Using GitHub Actions, you can define custom workflows that automatically build and test every pull request made to your repository or deploy merged pull requests to production.

GitHub Actions offers more than just DevOps functionality as it is closely integrated with GitHub. This allows you to run workflows in response to other repository events. For instance, you can have a workflow that adds relevant labels automatically when a new issue is created in your repository.

With GitHub Actions, you're in control. You can run your workflows using GitHub's Linux, Windows, and macOS virtual machines. You can also take full charge and operate self-hosted runners in your own data center or cloud infrastructure.

We will create a GitHub Action workflow to utilize the Linux agents hosted by GitHub.

## Preparation

We need to configure a few things before we can start working through our GitHub Action workflow code:

1.  Create a GitHub repository to host our code and workflow.
2.  Generate an SSH key pair; this will be used to access our Azure-hosted virtual machine instance from the GitHub-hosted compute resource when the workflow runs.
3.  Configure some repository secrets that will be used in our workflow; these will store things such as our Azure credentials and the SSH key pair we created.
4.  Copy the files from `https://github.com/PacktPublishing/Learn-Ansible-Second-Edition/tree/main/Chapter15` to your new repo and run the workflow.

Let's look at these steps in more detail.

### *Creating a repository*

Let's start by creating a repository in GitHub that we will use to host our code and workflow.

First, you need to log into GitHub. Once you've logged in, go to **Repositories** and then click on the **New** button; you will be taken to the **Create a new repository** page, where you need to update the following:

-   **Owner**: Here, you need to select an owner for the repository. This will typically be your GitHub user; however, if you belong to any organization, you may have the option of creating the repository under one of those organizations. If you do that, please ensure that you have permission to do so as we will be spinning up temporary compute resources, which may not be allowed by your organization's admins.

- **Repository name**: I recommend using something descriptive, such as *Learn-Ansible-Second-Edition-Chapter15*.

- **Description**: While this is optional, it is always best to add one; for example, let's add *Following along with Chapter 15 of Learn Ansible*.

- **Public** or **Private**: I recommend setting your repository's visibility to *private*.

You can leave the remaining options as-is and then click on the **Create repository** button at the end of the form. Once the repository has been created, you should be presented with a page that looks like this:

Figure 15.1 – Our new repository

Let's move on to the next step.

### Generating the SSH key pair and Azure Service Principle

We need to generate an SSH key pair and an Azure Service Principle before adding secrets to our newly created repository.

> **Information**
>
> Remember to run the commands in Windows Subsystem for Linux if you're following along on a Windows machine.

To do this, open a Terminal and run the following command:

```
$ ssh-keygen -t rsa -C "learnansible" -f ./id_rsa
```

When prompted to enter a passphrase, just hit *Enter*; we don't want to use one. This should give you two files: one called `id_rsa`, which contains the private portion of our key – please keep this private – and another called `id_rsa.pub`. As its name implies, it includes the public portion of our SSH key.

Next, we need to generate an Azure Service Principle and grant permissions to our Azure subscription.

From *Chapter 7*, *Ansible Windows Modules*, and *Chapter 9*, *Moving to the Cloud*, we used the Azure command-line tool to log in using our Azure credentials. However, when interacting with Azure using services such as GitHub Actions, we don't want to use our credentials as they will be locked down with multi-factor authentication, and you don't want to hand out your credentials.

To get around this, we can create a service principal and grant it permissions to the Azure subscription so that it can launch resources from the GitHub Action.

To create the service principle, you need to log into Azure using the Azure CLI by running the following command:

```
$ az login
```

If you are already logged in, run the following:

```
$ az account list
```

Both commands will return a list of subscription IDs your account can access. Please make a note of the ID; we will need it momentarily.

Here's an example of the sort of output you can expect to see; this is the JSON that is being returned by the API request that the Azure CLI has made:

```
{
  "environmentName": "AzureCloud",
  "id": "e80d5ad9-e2c5-4ade-a866-bcfbae2b8aea",
  "isDefault": true,
  "name": "My Subscription",
  "state": "Enabled",
  "tenantId": "c5df827f-a940-4d7c-b313-426cb3c6b1fe",
  "user": {
    "name": "account@russ.foo",
    "type": "user"
  }
}
```

The information we are after is labeled as `id` against the subscription to which we would like to grant the service principal access. Using the preceding example, the command I would need to run would be as follows:

```
$ az ad sp create-for-rbac -name sp-learn-ansible -role contributor -
scopes /subscriptions/e80d5ad9-e2c5-4ade-a866-bcfbae2b8aea
```

When you run this command, replace the subscription ID in the scope with your own.

The output you get will look something like this; please note it down as you will not be able to retrieve the password again:

```
Creating 'contributor' role assignment under scope '/subscriptions/
e80d5ad9-e2c5-4ade-a866-bcfbae2b8aea'
The output includes credentials that you must protect. Be sure
that you do not include these credentials in your code or check
the credentials into your source control. For more information, see
https://aka.ms/azadsp-cli
{
  "appId": "2616e3df-826d-4d9b-9152-3de141465a69",
  "displayName": "sp-learn-ansible",
  "password": "Y4j8Q~gVO*NoTaREalPa55w0rdpP-pdaw",
  "tenant": "c5df827f-a940-4d7c-b313-426cb3c6b1fe"
}
```

Also, as I am sure you will have already guessed, none of the information in the preceding examples is valid data, so please use your values in the next section.

### GitHub personal access token

There is one more set of credentials we need to generate; because our GitHub repository is set to private, we need to be able to authenticate to check the code out and write logs back to the repository during the workflow run. To do this, we will need to generate a personal access token.

A personal access token for GitHub is a secure, revocable, and customizable credential that allows you to authenticate with GitHub and access its API or command-line tools without using your main account password.

Rather than documenting the process here, as GitHub is moving from classic to fine-grained tokens at the time of writing, an up-to-date copy of the documentation can be found at https://docs.github.com/en/authentication/keeping-your-account-and-data-secure/managing-your-personal-access-tokens.

For our purposes, you need to name your token, select just your repository, and grant it the following access:

- **Contents**: Read-only
- **Metadata**: Read-only; this will be selected automatically once the permission is set

Once you have your token, please note it somewhere secure; it will not be displayed again.

## Adding secrets to the repository

Go back to the repository in GitHub by choosing **Settings | Secrets and Variables | Actions**. Click the **New repository secret** button for each of the secrets listed in the following table. Please make sure that you call each secret as per the following naming conventions since our workflow code references these secrets by their name:

| Secret Name | Secret Content |
| --- | --- |
| ARM_CLIENT_ID | This is the `appId` value from when you create the service principle. In this example, this would be `2616e3df-826d-4d9b-9152-3de141465a69`. |
| ARM_CLIENT_SECRET | This is the `password` value that was given when you created the service principle. In this example, this would be `Y4j8Q~gVO*NoTaREalPa55w0rdpP-pdaw`. |
| ARM_SUBSCRIP-TION_ID | This is your Azure subscription ID; use the one you granted the service principal access to. In this example, this would be `e80d5ad9-e2c5-4ade-a866-bcfbae2b8aea`. |
| ARM_TENANT_ID | This is the ID of the `tenant` value listed when you created the service principle. In this example, this would be `c5df827f-a940-4d7c-b313-426cb3c6b1fe`. |
| SSH_PRIVATE_KEY | Open the `id_rsa` file in a text editor and copy and paste the contents here. |
| SSH_PUBLIC_KEY | Open the `id_rsa.pub` file in a text editor and copy and paste the contents here. |
| GH_PAT | This should contain your GitHub personal access token. |

Table 15.1 – Information needed for GitHub Actions

Once they have all been added, your **Actions secrets and variables** page should look something like this:

Figure 15.2 – All of the repository secrets have been added

Now that we have all the basic configurations for the GitHub Action, let's look at the workflow itself.

## Understanding the GitHub Action workflow

The workflow file, which lives in the `.github/workflows/action.yml` file, contains, as its name suggests, the YAML code containing the jobs, steps, and tasks that will be executed during the workflow run. In our case, the workflow will execute the following two jobs, with each job being made up of multiple steps:

- Scan the Ansible Playbook:

  A.  Check out the code.

  B.  Create a folder to store the scan results.

  C.  Run a KICS scan on the checked-out code.

  D.  Upload a copy of the results to GitHub.

Now, if KICS detects a problem with our playbook, it will report an error, and the workflow will stop here – if everything looks good with the KICS scan, then the workflow will proceed by running the following job:

- Install and run the Ansible Playbook:

  A.  Check if a cached version of our Ansible modules and Python packages is available.

  B.  If not cached, download and install the Ansible Azure modules and the supporting Python packages.

  C.  Check out the code.

  D.  Log into Azure using the Azure CLI and the service principle we created.

  E.  Set the SSH key.

  F.  Run the Ansible Playbook, logging the output of the Playbook so that we can store a copy alongside the scan results in the workflow logs.

  G.  Upload the Playbook execution summary.

Now that we know what the workflow will do, let's dive into the code. We'll start with some basic configuration:

1.  The first line disables a KICS check – while the workflow does not form part of our Playbook, it is stored in the repository and will be scanned as part of the workflow's execution:

    ```
    # kics-scan disable=555ab8f9-2001-455e-a077-f2d0f41e2fb9
    name: "Ansible Playbook Run"
    env:
      FAIL_ON: "medium"
      RESULTS_DIR: "results-dir"
    ```

We are also setting the name of the workflow, which is how it will appear in the GitHub web interface, before finally setting up some variables that we will use during the workflow's execution.

2.  Next up, we have the configuration that defines the workflow that should run; for our needs, we will run the workflow each time the code is committed to the main branch:

```
On:
  push:
    branches:
      - main
```

3.  Next up, we must define our first job, which is the one that scans the Playbook code:

```
jobs:
  scan_ansible_playbook:
    name: "Scan Ansible Playbook"
    runs-on: ubuntu-latest
    defaults:
      run:
        shell: bash
```

As you can see, we are defining it as `scan_ansible_playbook`, which runs on the latest version of the Ubuntu image supplied by GitHub, and the default action for tasks is to run bash. With the job defined, we can move on to the next steps.

4.  We start with the ones that check out the code and create the directory where we are going to be storing the results of the scan we will be running:

```
    steps:
      - name: "Checkout the code"
        uses: "actions/checkout@v4"
        with:
          token: "${{secrets.GH_PAT}}"
```

5.  The step downloads a copy of the repository in which the workflow is hosted; as you can see, we are using `${{secrets.GH_PAT}}`. We will look at secret variables a little later. Now, we must create the folder:

```
      - name: "Create the folder for storing the scan results"
        run: mkdir -p ${{env.RESULTS_DIR}}
```

The section step creates a directory whose name is referenced as the `RESULTS_DIR` environment variable, which we defined in the top section of the workflow file.

6. When referencing an environment variable, we use the `${{env.VARIABLE_NAME}}` format. So, in our case, we are using `${{env.RESULTS_DIR}}`. In the next step, we have a dedicated task for running KICS, which is managed and maintained by Checkmarx:

```
- name: "Run kics Scan"
  uses: "checkmarx/kics-github-action@v1.7.0"
  with:
    path: "./"
    output_path: "${{env.RESULTS_DIR}}"
    output_formats: "json,sarif"
    fail_on: "${{ env.FAIL_ON }}"
    enable_jobs_summary: true
```

As you can see, we are instructing the task to output the JSON and SARIF files, **SARIF**, which stands for **Static Analysis Results Interchange Format**, is a standardized JSON-based file format for the output of static analysis tools that allows you to share and integrate analysis results between different tools and platforms. The results are outputted to the `${{env.RESULTS_DIR}}` directory we created in the previous step and also for the workflow fail if the results of the scan contain anything with a severity defined in `${{ env.FAIL_ON }}`. We set this to `medium` at the start of the workflow file.

7. Now that we have completed the scan, we can review the workflow code for the job that installs and runs Ansible. This is called `run_ansible_playbook`:

```
run_ansible_playbook:
  name: "Install Ansible and run Playbook"
  runs-on: ubuntu-latest
  needs: scan_ansible_playbook
  defaults:
    run:
      shell: bash
```

As you can see, the job is defined the same as the first job, with one exception: we have added a needs line with a value of `scan_ansible_playbook`. This instructs the job to only run once `scan_ansible_playbook` has completed with a successful status.

8. The step of the job checks for the presence of three folders; if they exist, a cached version of those folders will be used, meaning that once the workflow has been run once, subsequent executions will be much quicker as we don't have to install the Ansible Galaxy modules and their requirements each time the workflow runs:

```
steps:
  - name: "Cache Ansible collections and Python packages"
    uses: actions/cache@v4
    with:
      path: |
```

```
                 ~/.ansible/collections
                 ~/.cache/pip
                 /home/runner/.local/lib/python3.10/site-packages
            key: ${{ runner.os }}-ansible-collections-and-python-
packages
            restore-keys: |
              ${{ runner.os }}-ansible-collections-and-python-
packages
```

9. Next up, we have the step that checks out our repo:

```
    - name: "Checkout the code"
      id: "checkout"
      uses: "actions/checkout@v4"
```

You might be wondering, "*Why do we need to check out the code again? We already did that during the last job.*" This is a great question.

The answer is that the compute resource that ran the job was terminated when the last job finished running, and all data was lost. When the current job started, a new resource was launched, and we started again with a completely fresh installation.

10. The next step in the workflow uses the `Azure/login@2` task to install the Azure CLI if it's not already installed and then log in using the service principal information we defined as repository secrets earlier in this chapter:

```
    - name: "Login to Azure using a service principal"
      uses: "Azure/login@v2"
      with:
        creds: '{"clientId":"${{secrets.ARM_CLIENT_
ID }}","clientSecret":"${{secrets.ARM_CLIENT_SECRET
}}","subscriptionId":"${{secrets.ARM_SUBSCRIPTION_ID
}}","tenantId":"${{secrets.ARM_TENANT_ID }}"}'
```

We need to embed secrets using the `${{ secrets.SECRET_NAME }}` format. Here, we are using the following:

- `${{secrets.ARM_CLIENT_ID }}`

- `${{secrets.ARM_CLIENT_SECRET}}`

- `${{ secrets.ARM_SUBSCRIPTION_ID }}`

- `${{ secrets.ARM_TENANT_ID }}`

Because these are all defined as secrets, the values will never appear in any of the Pipeline run logs.

This means that while we know the values, someone else who has permission to run the workflow will never need to be told the credentials for our service principle as they can consume the secrets. They will also never accidentally be exposed to them if they check any logs or try and output them due to the workflow's execution as they will be automatically redacted.

11. The final step before we run Ansible is to add and configure the SSH key pair to our host:

```
- name: "Setup SSH key for Ansible"
  id: "add-ssh-key"
  run: |
    mkdir ~/.ssh
    chmod 700 ~/.ssh/
    echo "${{ secrets.SSH_PRIVATE_KEY }}" > ~/.ssh/id_rsa
    chmod 600 ~/.ssh/id_rsa
    echo "${{ secrets.SSH_PUBLIC_KEY }}" > ~/.ssh/id_rsa.pub
    chmod 644 ~/.ssh/id_rsa.pub
    cat  ~/.ssh/id_rsa.pub
```

12. The SSH key pair is the final piece we needed. Now, we can run Ansible:

```
- name: "Run the playbook (with ansible-playbook)"
  id: "ansible-playbook-run"
  continue-on-error: true
  run: |
    ansible-playbook -i inv site.yml 2>&1 | tee ansible_output.log
    echo "summary<<EOF" >> $GITHUB_OUTPUT
    echo "## Ansible Playbook Output" >> $GITHUB_OUTPUT
    echo "<details><summary>Click to expand</summary>" >> $GITHUB_OUTPUT
    echo "" >> $GITHUB_OUTPUT
    echo "\`\`\`" >> $GITHUB_OUTPUT
    cat ansible_output.log >> $GITHUB_OUTPUT
    echo "\`\`\`" >> $GITHUB_OUTPUT
    echo "</details>" >> $GITHUB_OUTPUT
    echo "EOF" >> $GITHUB_OUTPUT
  env:
    ANSIBLE_HOST_KEY_CHECKING: "False"
```

As you can see, there is slightly more to running Ansible here than we have been doing on our local machines. The reason we are running the Ansible playbook is to capture its output and format the output so that it can be displayed in the GitHub Actions job log.

Here's a breakdown of what's happening:

- **Name**: This step is named `Run the playbook (with ansible-playbook)` for clarity in the workflow's execution log.

- **ID**: The step is given an identifier of `ansible-playbook-run` so that we can refer to this step's outputs in the subsequent step.

- **Continue on Error**: By setting `continue-on-error` to `true`, we are allowing the workflow to continue even if this step encounters an error. This is useful for ensuring that the workflow can proceed to steps that might, for example, provide diagnostic information or perform cleanup actions, even if the Ansible playbook fails.

- **Run**: This key starts a multi-line script block that's executed in the jobs shell. The script does the following:

```
ansible-playbook -i inv site.yml 2>&1 | tee ansible_output.log
```

This command runs the Ansible playbook defined in `site.yml` with an inventory file, `inv`. The `2>&1` part redirects `stderr` to `stdout`, so both standard output and errors from the `ansible-playbook` command are piped to the `tee` command. `tee ansible_output.log` writes the output to `ansible_output.log` and displays it in the workflow's log for real-time monitoring.

Subsequent `echo` commands and `cat` append a formatted summary of the Ansible output to the special `GITHUB_OUTPUT` environment variable. As you may have noticed, we are mostly using Markdown to format the text.

- **Env**: The env section defines environment variables for this step. `ANSIBLE_HOST_KEY_CHECKING: "False"` disables Ansible's SSH host key checking. This option is often used in automated environments to avoid manual interventions.

The final step in our workflow takes the output of the previous step and outputs it to `$GITHUB_STEP_SUMMARY`. This is a special variable that's used by a GitHub Actions workflow to record the results of a step in the workflow executions log:

```
- name: "Publish Ansible Playbook run to Task Summary"
  env:
    SUMMARY: ${{ steps.ansible-playbook-run.outputs.summary }}
  run: |
    echo "$SUMMARY" >> $GITHUB_STEP_SUMMARY
```

While that completes our workflow code review, one more task happens in the background that we don't have to define. As you may recall, in the first step of the `run_ansible_playbook` job, we had a step that looked for any caches associated with the workflow. Well, by defining that step, there is a post-deploy task that runs at the end of the workflow and creates the cache if one doesn't exist.

Now that we understand our workflow code, let's check out a copy of our newly created repository. Copy the code from the example repository and then check in the changes.

## Committing the code

As mentioned previously, before running the workflow, we need to check out the empty repository we created at the start of this chapter. This will vary depending on how you interact with GitHub. I use the command line, but you might use the GitHub Desktop application or an IDE such as Visual Studio Code.

> **Information**
>
> For more information on the GitHub desktop application, see `https://desktop.github.com/`. For details on how to configure an SSH connection to GitHub, see `https://docs.github.com/en/authentication/connecting-to-github-with-ssh`.

If you want to follow along on the command line, you must update the repository's name to reflect your own and ensure you have SSH access to your GitHub repositories:

```
$ git clone https://github.com/PacktPublishing/Learn-Ansible-Second-
Edition.git
$ cd Learn-Ansible-Second-Edition-Chapter15
```

Once I was in the folder, I copied across the contents of `https://github.com/PacktPublishing/Learn-Ansible-Second-Edition/tree/main/Chapter15`, ensuring that I also copied the `.github` folder as this contains the workflow we want to execute.

Once copied, I ran the following commands to add the new files and create the first commit, then pushed:

```
$ git add .
$ git commit -m "first commit"
$ git push
```

If everything goes as planned, if you go to your repository and click on the **Actions** tab, you should see something like this:

Figure 15.3 – Our first commit is running the GitHub Action

Clicking the name of the commit should show you the progress of the workflow:

Figure 15.4 – Viewing the progress of the workflow

Click on the running job – in my example, this is the *Install Ansible and Run Playbook* job. This will show you its real-time progress:

Figure 15.5 – Viewing the real-time output

If everything works as planned, the Ansible playbook will run, the Azure resources will be deployed, and we should have a running WordPress instance.

Clicking on the **Summary** link at the top of the page will show you the full output. Here, we'll see any warnings or information that was logged during the workflow run, followed by the KICS results:

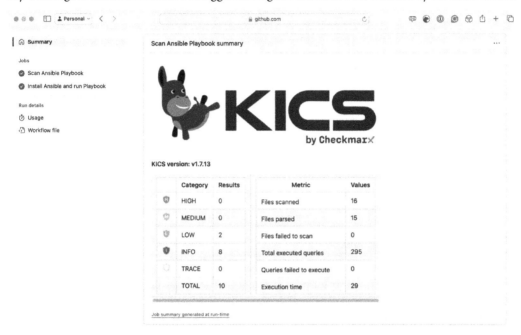

Figure 15.6 – The KICS scan results

You will also be able to expand the **Ansible Playbook Output** area and view the logs:

Figure 15.7 – Ansible Playbook Output

Before we remove the Azure resources, let's see what happens when the scan fails. To do this, open `roles/azure/tasks/main.yml` and remove the line that reads as follows (it should be around line *61*):

```
security_group: "{{ nsg_output.state.name }}"
```

Once removed, check in the updated code. This will trigger a new workflow run:

Figure 15.8 – Triggering a second workflow run

As the line we removed will trigger a medium severity rule, our workflow run should fail, as shown here:

Figure 15.9 – Our second workflow run failed due to our change

Once you have finished testing, I recommend logging into Azure and manually deleting the resource group containing the resources we've just launched.

As you can see, while there are considerations you need to make for your deployments – such as ensuring all the connectivity and steps are in place to interact with your cloud provider securely – the general gist and approach to running our Playbooks remains much the same as on our local machine.

The same can also be said about the next tool we will examine, Azure DevOps.

# Azure DevOps

The description we used for GitHub Actions also applies to Azure DevOps Pipelines and repositories, two of the Azure DevOps services we will use in this section. Again, we will use platform-provided computing resources to run our Ansible Playbook, and many approaches will be the same. So, rather than covering old ground, let's start with preparing an Azure DevOps project to host our code and run our Playbook.

## Creating and configuring our project

First, you will need to create an Azure DevOps project. Like our GitHub repository, I've called it `Learn-Ansible-Second-Edition-Chapter15`:

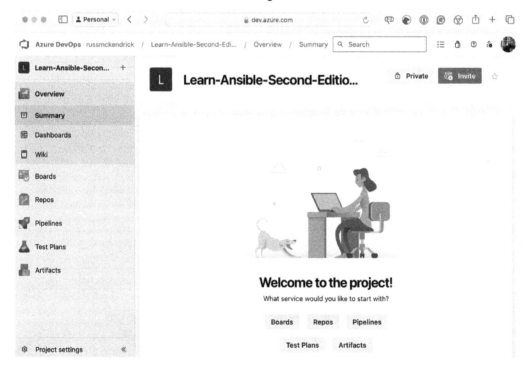

Figure 15.10 – Our newly created Azure DevOps project

We need to configure a few things before checking our code in and adding our pipeline; the first is to create a service connection to Azure itself. To do this, click on the **Project Settings** button, which can be found at the bottom far left-hand corner of the page.

Once **Project Settings** is open, in the left-hand menu under **Pipelines**, click **Service connections**, then click the **Create service connection** button.

Select **Azure Resource Manager**, then click **Next**; from here, select **Service principal (manual)** and click **Next** again.

We are taking this approach rather than any of the others, which would automatically create the service principle for us, as we already have the details of a service principle noted from the *GitHub Actions* section.

The following table contains the information you need to enter:

| Option | Content |
| --- | --- |
| Subscription Id | This is your Azure subscription ID; use the one you granted the service principal access to. In this example, this would be `e80d5ad9-e2c5-4ade-a866-bcfbae2b8aea`. |
| Subscription Name | Enter the name of your Azure subscription. Since we are going to be referring to the subscription ID in the pipeline code, this can be set to anything you like. |
| Service Principal Id | This is the `appId` value from when you create the service principle. In this example, this would be `2616e3df-826d-4d9b-9152-3de141465a69`. |
| Service principal key | This is the `password` value that was given when you created the service principle. In this example, this would be `Y4j8Q~gVO*NoTaREal-Pa55w0rdpP-pdaw`. |
| Tenant ID | This is the ID of the `tenant` value listed when you created the service principle. In this example, this would be `c5df827f-a940-4d7c-b313-426cb3c6b1fe`. |
| Service connection name | Enter `azConnection` here as this is how we are referencing the connection in the pipeline code. |
| Security | Ensure that Grant access permission to all pipelines is selected. |

Table 15.2 – Information needed for your pipeline in Azure DevOps

Once you've entered this information, click the **Verify and Save** button. This will check the details you entered are correct and save the service connection.

Next, we need to install a few extensions from the Visual Studio Marketplace, allowing us to publish our KICS report and an overview of our Playbook run:

- **Markdown Reports**: `https://marketplace.visualstudio.com/items?itemName=MasamitsuMurase.publish-markdown-reports`

- **Sarif Tools**: `https://marketplace.visualstudio.com/items?itemName=sariftools.scans`

To enable the extensions on your Azure DevOps organization, follow the preceding URLs and the instructions when you click the **Get it free** button.

The final configuration piece is adding a pipeline variable group and secure file. To do this, click on **Pipelines** in the left-hand side menu and then click **Library**. Once on the **Library** page, click the + **Variable group** button.

Name the variable group `playbook` and enter the following variables:

| Name | Value |
| --- | --- |
| `breakSeverity` | `MEDIUM` |
| `SSH_PUBLIC_KEY` | Paste the contents of the `id_rsa.pub` file here |
| `subscriptionName` | `azConnection` – this is the name of the connection we created at the start of this section |

Table 15.3 – Information required for the variable group

Once you have filled in the preceding information, click **Save**. Once saved, return to the pipeline **Library** area and click **Secure files**; once there, click the + **Secure file** button and upload the `id_rsa` file.

We now have all the base configurations ready and can upload our code.

## Cloning the repository and uploading the code

Next, we must clone the repository and upload our code, including the `azure-pipelines.yml` file we will cover in the next section. To do this, click on **Repos** in the left-hand side menu; you will be presented with several ways to clone the repository.

I've chosen to clone using SSH again; if you are following along, update the `git clone` command to reflect your repository:

```
$ git clone git@ssh.dev.azure.com:v3/russmckendrick/Learn-Ansible-
Second-Edition-Chapter15/Learn-Ansible-Second-Edition-Chapter15
$ cd Learn-Ansible-Second-Edition-Chapter15
```

I then copied the files across from https://github.com/PacktPublishing/Learn-Ansible-Second-Edition/tree/main/Chapter15. This time, I didn't worry about copying the .github directory as it isn't required. Once the files were in my locally cloned folder, I ran the following commands to add the new files and create the first commit, then push:

```
$ git add .
$ git commit -m "first commit"
$ git push
```

Unlike when we first checked our code into GitHub, nothing will happen because we haven't configured our pipeline yet.

## The Azure DevOps pipeline

Our pipeline is defined in the azure-pipelines.yml file, which can be found at the root of our repository file. Let's quickly review the content before we create the pipeline using that file.

> **Information**
>
> Structurally, our azure-pipelines.yml file is close to what we have already covered for GitHub Actions; in fact, you might almost think they are interchangeable and compatible – however, they aren't, so please be careful not to mix the two up.

Our pipeline file starts with a basic configuration that instructs the pipeline when to trigger, which variable group to load, and which underlying image to use. Right at the top, there's an exclusion rule for KICS, something we covered in *Chapter 13*, *Scanning Your Ansible Playbooks*:

```
# kics-scan disable=3e2d3b2f-c22a-4df1-9cc6-a7a0aebb0c99
trigger:
  - main
variables:
  - group: playbook
pool:
  vmImage: ubuntu-latest
```

Once the basic configuration is complete, we can start the stages:

1.  Our first run is the KICS scan on the code:

    ```
    - stage: "scan"
      displayName: "KICS - Scan Ansible Playbook"
    ```

2.  This stage is made up of a single job:

    ```
    jobs:
      - job: "kics_scan"
        displayName: "Run KICS Scan"
        pool:
          vmImage: "ubuntu-latest"
        container: checkmarx/kics:debian
    ```

3.  As you may have noticed, here, we are using the `checkmarx/kics:debian` container image to deploy KICS. This will spin up the container and run the following steps from within it. Our step contains two tasks – the first creates the output folder, checks out the code, and runs the scan:

    ```
    steps:
      - script: |
          mkdir -p $(System.DefaultWorkingDirectory)/output
          /app/bin/kics scan --ci -p ${PWD} -o ${PWD}
    --report-formats "all" --ignore-on-exit results
          mv results* $(System.DefaultWorkingDirectory)/
    output
          ls -lhat $(System.DefaultWorkingDirectory)/output
    ```

4.  The second task publishes the content of the output directory, which contains all of our scan results as a build artifact:

    ```
      - task: PublishBuildArtifacts@1
        inputs:
          pathToPublish: $(System.DefaultWorkingDirectory)/
    output
          artifactName: CodeAnalysisLogs
    ```

5.  With the files published, we no longer need the resources that were generated during this stage, so we can move on to the second stage:

    ```
    - stage: "scan_parse"
      displayName: "KICS - Parse Scan Resaults"
      jobs:
        - job: "kics_scan_parse_result"
          displayName: "Check KICS Scan Resaults"
          pool:
            vmImage: "ubuntu-latest"
          steps:
    ```

6. As you can see, this stage parses our scan results; the first task we run downloads a copy of the artifact we uploaded during the last stage:

```
- task: DownloadPipelineArtifact@2
  displayName: "Download the Security Scan Artifact
Result"
  inputs:
    artifact: CodeAnalysisLogs
```

Now that we have the results files, we need to review them to figure out if the Ansible Playbook should be run or not. This task runs a bash script that reads the JSON results and sets some pipeline variables to control what happens next.

7. We start the task with some configuration:

```
- task: Bash@3
  name: "setvar"
  displayName: "Check for issues in the scan result"
  inputs:
    failOnStderr: true
    targetType: "inline"
    script: |
```

8. Now, we have the script itself, which starts by setting some local variables and printing some results out to the screen using the `echo` command. These will appear in our pipeline run:

```
resultsFilePath="$(Pipeline.Workspace)/
results.json"
BREAK=$(breakSeverity)
echo "Checking for severity level: $BREAK"
noIssues=$(jq --arg BREAK "$BREAK" '.severity_
counters[$BREAK] // 0' $resultsFilePath)
echo "Number of issues found: $noIssues"
```

Then, we create a *group*, which means that when we review the pipeline output, the following information will be minimized, making it easier to read.

9. In the group, we have an `if` statement that states that if less than (`-lt`) 1 issues are detected (that is, zero issues), then the output variable, OK_TO_DEPLOY, is set to `true`:

```
echo "##[group]Checking the scan output"
if [ "$noIssues" -lt 1 ]; then
    echo "##vso[task.setvariable variable=OK_
TO_DEPLOY;isOutput=true]true"
    echo "##vso[task.logissue type=warning]No
issue found. Progressing with pipeline."
```

10. If this condition is not met – that is, there are one or more issues – then OK_TO_DEPLOY is set to `false` and an error is logged:

```
            else
                echo "##vso[task.setvariable variable=OK_
TO_DEPLOY;isOutput=true]false"
                echo "##vso[task.logissue type=error]
Pipeline failed due to $noIssues issue(s) found."
            fi
            echo "##[endgroup]"
```

11. Logging the error will stop the remainder of the pipeline from running. The next and final stage runs the Ansible Playbook. It has a dependency on the previous stage being successfully executed and OK_TO_DEPLOY being set to `true`:

```
  - stage: "run_ansible"
    displayName: "Run Ansible"
    condition: |
      and
      (
        succeeded(),
        eq(dependencies.scan_parse.outputs['kics_scan_parse_
result.setvar.OK_TO_DEPLOY'], 'true')
      )
    jobs:
      - job: "ansible_install"
        displayName: "Ansible"
        steps:
```

12. The first task logs us into Azure and sets the service principle details as environment variables for use in a later task:

```
        - task: AzureCLI@2
          displayName: 'Azure CLI'
          inputs:
            azureSubscription: '$(subscriptionName)'
            addSpnToEnvironment: true
            scriptType: 'bash'
            scriptLocation: 'inlineScript'
            inlineScript: |
              echo "##vso[task.setvariable variable=ARM_
SUBSCRIPTION_ID]$(az account show --query="id" -o tsv)"
              echo "##vso[task.setvariable variable=ARM_
CLIENT_ID]${servicePrincipalId}"
```

```
            echo "##vso[task.setvariable variable=ARM_
    CLIENT_SECRET]${servicePrincipalKey}"
            echo "##vso[task.setvariable variable=ARM_
    TENANT_ID]${tenantId}"
```

13. Next up, we need to add our SSH key to our environment. This uses the secure file we uploaded earlier:

```
- task: InstallSSHKey@0
  displayName: "Add SSH Key"
  inputs:
    sshKeySecureFile: "id_rsa"
    knownHostsEntry: "azure.devops"
```

14. Now, we need to add the public portion of the SSH key, install the bits we need to run the Ansible Playbook, and then actually run it, remembering to add the details for the service principle:

```
- task: Bash@3
  name: "ansible"
  displayName: "Run Ansible"
  env:
    AZURE_CLIENT_ID: $(ARM_CLIENT_ID)
    AZURE_SECRET: $(ARM_CLIENT_SECRET)
    AZURE_TENANT: $(ARM_TENANT_ID)
    AZURE_SUBSCRIPTION_ID: $(ARM_SUBSCRIPTION_ID)
    ANSIBLE_HOST_KEY_CHECKING: "False"
  inputs:
    targetType: "inline"
    script: |
```

15. With the environment ready, we can run the script, which starts by adding the `id_rsa.pub` file and adding the right permissions:

```
        echo "##[group]Add SSH key"
            echo "$(SSH_PUBLIC_KEY)" > ~/.ssh/id_rsa.
pub
            chmod 644 ~/.ssh/id_rsa.pub
        echo "##[endgroup]"
```

16. The next part of the script installs the Azure Ansible collection from Ansible Galaxy and installs the requirements. We are using `--force` here to ensure that the latest copy of all the collection is pulled down from Ansible Galaxy:

```
        echo "##[group]Install the Azure Ansible
Collection"
            ansible-galaxy collection install --force
azure.azcollection
```

```
                    pip3 install -r ~/.ansible/collections/
        ansible_collections/azure/azcollection/requirements-azure.txt
                    echo "##[endgroup]"
```

17. With those installed, we can now run the playbook; we are taking a similar approach to running the playbook as we did for our GitHub Action:

```
            echo "##[group]Run the Ansible Playbook"
                    ansible-playbook -i inv site.yml 2>&1 |
        tee $(System.DefaultWorkingDirectory)/ansible_output.log
            echo "##[endgroup]"
```

18. The final part of our script takes our Ansible output and creates a Markdown file called summary. md:

```
                    echo "##[group]Create the mardown file for the
        Ansible Playbook Output"
                    mkdir -p $(System.
        DefaultWorkingDirectory)/markdown
                    echo "# Ansible Playbook Output" >
        $(System.DefaultWorkingDirectory)/markdown/summary.md
                    echo "<details><summary>Click to expand</
        summary>" >> $(System.DefaultWorkingDirectory)/markdown/summary.
        md
                    echo "" >> $(System.
        DefaultWorkingDirectory)/markdown/summary.md
                    echo "\`\`\`" >> $(System.
        DefaultWorkingDirectory)/markdown/summary.md
                    cat $(System.DefaultWorkingDirectory)/
        ansible_output.log >> $(System.DefaultWorkingDirectory)/
        markdown/summary.md
                    echo "\`\`\`" >> $(System.
        DefaultWorkingDirectory)/markdown/summary.
        md                    echo "</details>" >> $(System.
        DefaultWorkingDirectory)/markdown/summary.md
                    echo "##[endgroup]"
```

19. The final task of the pipeline is to upload a copy of the markdown/summary.md file to our pipeline:

```
        - task: PublishMarkdownReports@1
          name: "upload_ansible_output"
          displayName: "Upload Ansible Output"
          inputs:
            contentPath: "$(Build.SourcesDirectory)/markdown"
            indexFile: "summary.md"
```

With that, our pipeline is complete. So, now that we know what it does, let's add it to our Azure DevOps project and run it for the first time.

If you click on **Pipelines** in the left-hand side menu and then click the **Create Pipeline** button, you will be asked, **Where is your code?**. select **Azure Repos Git**, and then your repository – the `azure-pipelines.yml` file will be loaded and you will have the option to **Run** or **Save**. We'll click **Run**.

You will be presented with something like the following screen:

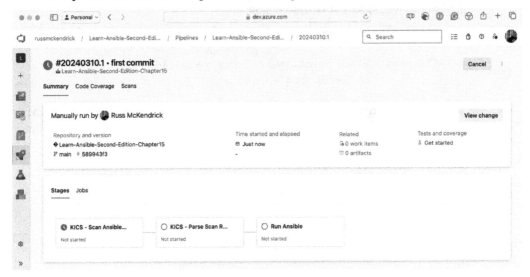

Figure 15.11 – Running the pipeline for the first time

However, not is all as it seems! If you click on the first stage, you will be presented with the following. The pipeline needs permissions to access the variable group we created:

Figure 15.12 – Granting the permissions for the variable group

Click **View** and follow the onscreen instructions to grant the permissions. The KICS scan will run, and the stage will be complete. It will then move on to the **Parse Scan Results** stage, which should be completed again.

If you go back to the summary, you'll see that more permissions are required, this time to access the secure file we uploaded:

Figure 15.13 – Grant the permissions for the secure file

Again, click **View** and follow the onscreen instructions to grant permission. This should be the last permission that needs to be given. From now on, when we run the pipeline, permissions will already be given.

If you click on the **Run Ansible** stage, you can keep track of the Playbook run. If everything goes as planned, returning to the summary should show you something like the following:

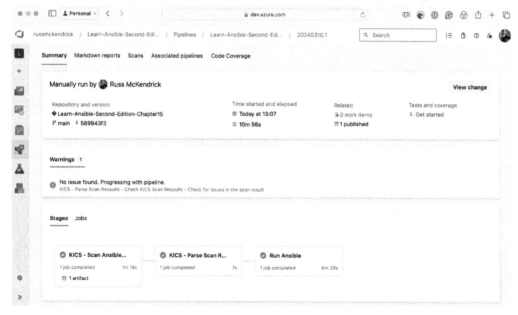

Figure 15.14 – Everything worked!!!

Clicking on **Markdown reports** will show the result of the Playbook run:

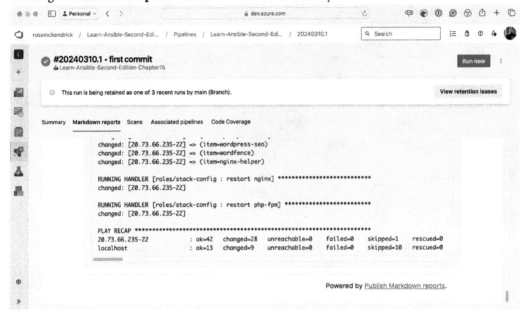

Figure 15.15 – The Markdown report

Clicking **Scans** will show you the results of the KICS scan:

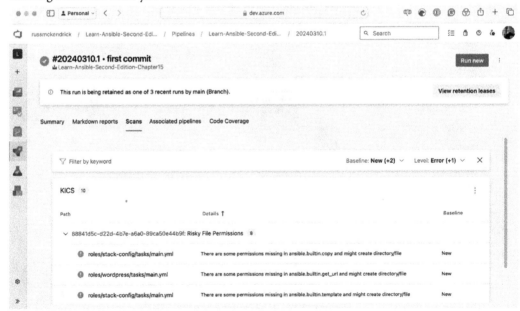

Figure 15.16 – The scan report

Like GitHub Actions, let's see what happens when the scan fails. Again, open `roles/azure/tasks/`
`main.yml` and remove the line that reads as follows (it should be around line *61*):

```
security_group: "{{ nsg_output.state.name }}"
```

Once removed, check in the updated code. This will trigger a new workflow run:

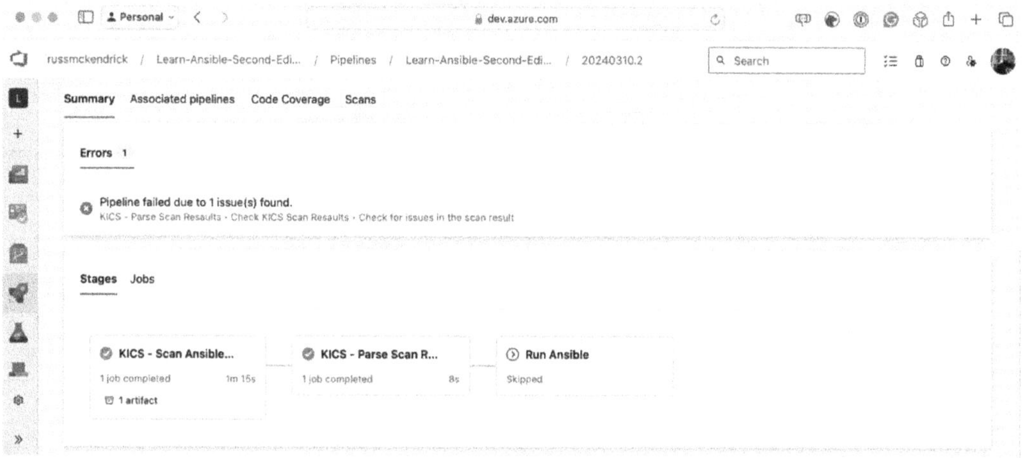

Figure 15.17 – The pipeline has errored

As you can see, we have a message stating **Pipeline failed due to 1 issue(s) found**, and the **Run Ansible**
stage was skipped as we didn't meet the conditions for it to run.

Once you have finished testing, log into Azure and manually delete the resource group containing
the resources we have just launched.

# Summary

In this chapter, we looked at running our Ansible Playbooks using the compute resources GitHub
and Azure DevOps provide. We discovered that this is great for running our playbook code as we can
ship code that defines the configuration for the computing resources alongside our Playbook code.

We also learned that by using the built-in tools, we can securely configure our environment so as
not to expose secrets, such as our service principle credentials, to other users who have access to run
the playbook.

The only downside is that we had to create the logic that runs the playbook. Wouldn't it be great to use
a tool designed to centrally run our Playbooks from a single user interface? Well, in our next chapter,
we will cover exactly that – so if you like the approach we have taken so far, read on.

# Further reading

To learn more about the topics that were covered in this chapter, take a look at the following resources:

- **GitHub Actions**: `https://docs.github.com/en/actions`
- **Azure DevOps**: `https://azure.microsoft.com/en-gb/products/devops`

# 16

# Introducing Ansible AWX and Red Hat Ansible Automation Platform

This chapter will examine two graphical interfaces for Ansible: the commercial **Red Hat Ansible Automation Platform** and the open source **Ansible AWX** – or to give it its full name, **Ansible Web eXecutable**.

This chapter will focus on the open source Ansible AWX because it is freely available and, outside of the resources required to run the tool, requires no upfront costs or contracts.

We will discuss how to install Ansible AWX and why you would want to use it. After all, we are 16 chapters into our journey with Ansible and haven't needed to use a graphical interface yet – so why now?

By the end of this chapter, we will have done the following:

- Discussed Red Hat Ansible Automation Platform versus Ansible AWX
- Installed and configured Ansible AWX
- Deployed our Microsoft Azure cloud application using Ansible AWX

## Technical requirements

While we will only deploy Ansible AWX in this chapter, its requirements are complex. Because of this, rather than running it locally, I will provide instructions for deploying a Kubernetes cluster in Microsoft Azure using the AKS service.

If you are following along, you will need access to a Microsoft Azure account and have the Azure CLI installed. For more information, see *Chapter 9, Moving to the Cloud*.

# Red Hat Ansible Automation Platform versus AWX

Red Hat Ansible Automation Platform and Ansible AWX are two powerful tools Red Hat provides for managing and streamlining your Ansible deployments. Both tools offer web-based interfaces that simplify the execution and management of Ansible playbooks, making it easier for users to leverage Ansible's automation capabilities without requiring extensive command-line knowledge.

Red Hat Ansible Automation Platform, formerly known as **Ansible Tower,** is a comprehensive enterprise-grade solution that goes beyond the capabilities of Ansible Tower. It integrates various components to create a cohesive and expansive automation environment. Some key features of Red Hat Ansible Automation Platform are as follows:

- **Centralized control**: Red Hat Ansible Automation Platform provides a unified web-based dashboard for defining, scheduling, and monitoring automation jobs from a central location.

- **Role-based access control (RBAC)**: With granular access, you can ensure that your users have appropriate access to automation resources, enhancing security and control.

- **Workflow management**: Create complex workflows that combine multiple playbooks, job templates, and inventory sources, as well as supporting dependencies, conditionals, and approvals.

- **Scalability and flexibility**: Automation can be scaled to meet the needs of large enterprises, supporting diverse infrastructures, including cloud platforms, containers, and network devices.

- **Content collections**: Access pre-packaged modules and plugins that have been expertly curated to expedite the implementation of automation projects.

- **Automation Hub**: This centralized repository hosts certified, partner-supported, and community-driven content. It fosters collaboration and accessibility to high-quality resources.

- **Automation analytics**: Utilize sophisticated analytics tools to scrutinize performance, utilization, and various KPIs across different clusters and instances.

- **Integration with Red Hat ecosystem**: Seamless integration with other Red Hat products such as Red Hat Insights and Red Hat Satellite, fostering a cohesive environment.

On the other hand, Ansible AWX is the open source upstream project for Red Hat Ansible Automation Platform. It provides many of the platform's core features but follows a community-driven development model with more frequent releases. While Ansible AWX offers a solid foundation for automation, Red Hat Ansible Automation Platform may need some enterprise-specific features and integrations.

The choice between Red Hat Ansible Automation Platform and Ansible AWX depends on your organization's needs and requirements. Red Hat Ansible Automation Platform is ideal for enterprises seeking a robust, feature-rich solution with commercial support and seamless integration with the Red Hat ecosystem. It offers advanced features and is designed to handle complex automation needs across diverse environments.

On the other hand, Ansible AWX is a suitable choice for organizations that prefer an open source solution and are comfortable with community-driven support. It provides a solid foundation for automation and benefits from more frequent updates and community contributions.

Both Red Hat Ansible Automation Platform and Ansible AWX allow organizations to automate at scale, reduce manual efforts, and improve the consistency and reliability of their IT operations. They provide user-friendly interfaces and enable effective team collaboration, increasing efficiency and improved compliance.

# Ansible AWX

To say that installing Ansible AWX is complicated is an understatement. Since Red Hat first open sourced the project, deploying it has always been difficult.

Luckily, the first release was containerized, and it has slowly transitioned from running in a small number of containers to being able to run in a Kubernetes cluster and managed by the AWX Operator.

> **Information**
>
> A Kubernetes Operator uses custom resources to automate application and component management in Kubernetes clusters. It extends the cluster's behavior without modifying the Kubernetes code itself. Operators can handle various tasks, such as deployment, backups, upgrades, and service discovery, reducing manual intervention and increasing the system's reliability.

Let's start by launching our own Kubernetes in Microsoft Azure and configuring our local machine so that we can deploy and configure the AWX Operator.

## Deploying and configuring the Ansible AWX Operator

The first thing we need to do is deploy the Kubernetes cluster. To do this, we will use the Azure CLI to launch an AKS cluster. To start with, we need to set some variables on the command line to define the resource names, which Azure region we would like the cluster to deploy into, and how many compute nodes we require:

```
$ AKSLOCATION=uksouth
$ AKSRG=rg-awx-cluster
$ AKSCLUSTER=aks-awx-cluster
$ AKSNUMNODES=2
```

Next up, let's create the Azure Resource Group we will be deploying our cluster into; this will make it easy to remove once we have finished as we need to delete the group and its contents:

```
$ az group create --name $AKSRG --location $AKSLOCATION
```

With the resource group in place, we can now launch the AKS cluster:

```
$ az aks create \
    --resource-group $AKSRG \
    --name $AKSCLUSTER \
    --node-count $AKSNUMNODES \
    --generate-ssh-keys
```

This will take around 5 minutes to deploy. If you don't have the kubectl command installed on your local machine, then you can run the following command to have the Azure CLI install it for you:

```
$ az aks install-cli
```

Finally, with kubectl installed, you can configure the credentials and contexts by running the following command:

```
$ az aks get-credentials --resource-group $AKSRG --name $AKSCLUSTER
```

With our cluster now launched and available, we must install and configure the AWX Operator using Helm.

> **Information**
>
> **Helm** is a package manager that simplifies Kubernetes deployment by packaging applications as charts and defining necessary resources and configurations. For more details and installation instructions, see https://helm.sh/.

First, we need to enable the AWX repository and pull it down to our local machine:

```
$ helm repo add awx-operator https://ansible.github.io/awx-operator/
$ helm repo update
```

Now, we need to deploy the AWX Operator to our cluster:

```
$ helm install -n awx --create-namespace awx awx-operator/awx-operator
--version 2.12.1
```

It will take a minute or two to deploy.

> **Please note**
>
> You might have noticed that the preceding command specifies an explicit version number because there are some known bugs with the current release, which is a major update from the version we are using.

You can run the following command to check the status of the deployment:

```
$ kubectl get pods -n awx
```

Once everything is ready, you should see something like the following screen:

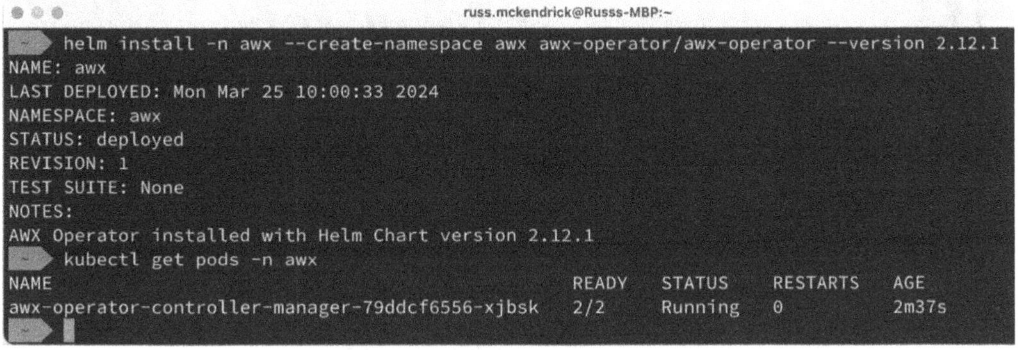

```
russ.mckendrick@Russ-MBP:~

     helm install -n awx --create-namespace awx awx-operator/awx-operator --version 2.12.1
NAME: awx
LAST DEPLOYED: Mon Mar 25 10:00:33 2024
NAMESPACE: awx
STATUS: deployed
REVISION: 1
TEST SUITE: None
NOTES:
AWX Operator installed with Helm Chart version 2.12.1
     kubectl get pods -n awx
NAME                                                  READY   STATUS    RESTARTS   AGE
awx-operator-controller-manager-79ddcf6556-xjbsk      2/2     Running   0          2m37s
```

Figure 16.1 – Deploying the AWX Operator

With the AWX Operator deployed with our cluster, we can request that the operator now deploy AWX itself. To do this, run the following command:

```
$ kubectl apply -f https://raw.githubusercontent.com/PacktPublishing/
Learn-Ansible-Second-Edition/main/Chapter16/awx/ansible-awx.yaml
```

This command simply passes the following YAML configuration to the operator to instruct it how to deploy our AWX installation:

```
---
apiVersion: awx.ansible.com/v1beta1
kind: AWX
metadata:
  name: ansible-awx
  namespace: awx
spec:
  service_type: loadbalancer
```

As you can see, there's not much to it, so please don't consider this a production-ready AWX instance. All we are instructing the AWX Operator to do is deploy AWX and expose the service via a load balancer so that we can connect to it.

Now, we wait; our AWX installation will take 15 to 20 minutes to deploy the application and bootstrap itself.

You can check the status of the containers and the load balancer service by running the following code:

```
$ kubectl get pods -n awx
$ kubectl get svc ansible-awx-service -n awx
```

Once the basics have been deployed, you should see something like the following. These are the containers that service the AWX application. As you can see, there are ones for the database, task runner, and the web interface:

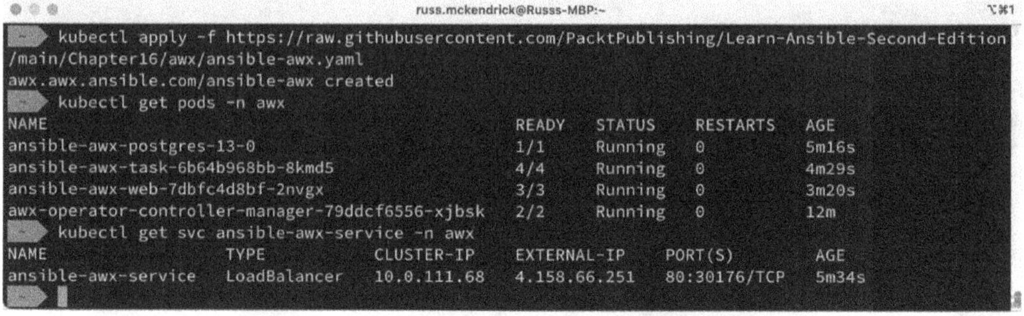

```
● ● ●                                    russ.mckendrick@Russs-MBP:~                              ⌥⌘1
   kubectl apply -f https://raw.githubusercontent.com/PacktPublishing/Learn-Ansible-Second-Edition
/main/Chapter16/awx/ansible-awx.yaml
awx.awx.ansible.com/ansible-awx created
   kubectl get pods -n awx
NAME                                                    READY   STATUS    RESTARTS   AGE
ansible-awx-postgres-13-0                               1/1     Running   0          5m16s
ansible-awx-task-6b64b968bb-8kmd5                       4/4     Running   0          4m29s
ansible-awx-web-7dbfc4d8bf-2nvgx                        3/3     Running   0          3m20s
awx-operator-controller-manager-79ddcf6556-xjbsk        2/2     Running   0          12m
   kubectl get svc ansible-awx-service -n awx
NAME                  TYPE           CLUSTER-IP      EXTERNAL-IP    PORT(S)        AGE
ansible-awx-service   LoadBalancer   10.0.111.68     4.158.66.251   80:30176/TCP   5m34s
```

Figure 16.2 – Checking the status of our AWX deployment

Once your deployment looks like the preceding output, the final step is to grab the admin password. To do this, run the following command – the secret will always be named `ansible-awx-admin-password`:

```
$ kubectl get secret -n awx ansible-awx-admin-password -o jsonpath="{.
data.password}" | base64 -decode
```

This will grab the base64 encoded secret from the Kubernetes secret store and decode it for you – it should look like this:

```
● ● ●                                    russ.mckendrick@Russs-MBP:~                              ⌥⌘1
   kubectl get secret -n awx ansible-awx-admin-password -o jsonpath="{.data.password}" | base64 --
decode
h6VBBzcnDTHiBbl7jZOmA30tpsjka8nF%
```

Figure 16.3 – Grabbing the admin password

As you may have noticed in the preceding output, there is a % icon at the end – this is not part of the password, and you need everything before that.

Please make a note of the password and the EXTERNAL-IP value from the previous commands as this tells you where to go to log in and what credentials to use. In the preceding deployment (which has long since been terminated), these details are as follows:

- **URL**: http://4.158.66.251/
- **Username**: admin
- **Password**: h6VBBzcnDTHiBbl7jZOmA30tpsjka8nF

When you go to the URL, you should be greeted with a login page that looks like this:

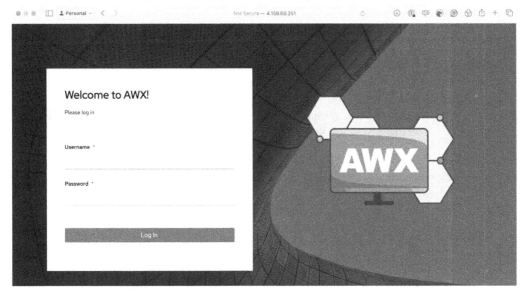

Figure 16.4 – Grabbing the admin password

Once you log in, you will be taken to your empty AWX instance:

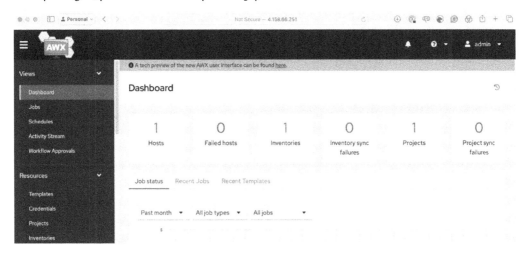

Figure 16.5 – Grabbing the admin password

Now, let's set up our playbook.

## Setting up our playbook

Before running our playbook, we must import it into Ansible AWX and configure the supporting credentials, such as our Azure Service Principle. We'll start with a project.

### Adding a new project

First, we need to add a new project, where we tell Ansible AWX about the repository hosting our playbook. As mentioned previously, we will use a GitHub repository that houses the code. To add a new project, click on **Projects** under **Resources** in the left menu and then click on the **Add** button.

Here, you will be asked for several bits of information; enter the following:

- **Name**: Azure WordPress
- **Description**: Deploy WordPress in Azure
- **Organization**: Default
- **Execution Environment**: Select **AWX EE (latest)**
- **Source Control Type**: GIT

When you select the **Source Control Type** value a second section will appear that asks for details about where your source is hosted:

- **Source Control URL**: https://github.com/PacktPublishing/Learn-Ansible-Second-Edition.git
- **Source Control Branch/Tag/Commit**: Leave blank
- **Source Control Refspec**: Leave blank
- **Source Control Credential**: Leave blank
- **Options**: Just select Clean

Once you have entered these details, click **Save**. Now, if you return to the **Projects** page, you should see that Ansible has already downloaded the source for the playbook:

Figure 16.6 – Adding the project and downloading the code from GitHub

## Adding credentials

Next, we must tell Ansible AWX the credentials to use when accessing our Azure environment; to add these, click **Credentials**. This can also be found under the **Resources** section of the left-hand menu. Click **Add** and enter the following:

- **Name**: `Azure`

- **Description**: `Credentials for Azure`

- **Organization**: `Default`

- **Credential Type**: Select **Microsoft Azure Resource Manager**

As before, this will open a separate section; here, you will need to enter details of the service principle we created in *Chapter 15*, *Using Ansible with GitHub Actions and Azure DevOps*:

- **Subscription ID**: Enter the subscription ID; in the example from the previous chapter, this was `e80d5ad9-e2c5-4ade-a866-bcfbae2b8aea`

- **Username**: Leave blank

- **Password**: Leave blank

- **Client ID**: Enter the `appId` value that was returned when you created the service principle; in the previous chapter's example, this was `2616e3df-826d-4d9b-9152-3de141465a69`

- **Client Secret**: Enter the `password` value that was returned when you created the service principle; in the previous chapter's example, this was `Y4j8Q~gVO*NoTaREalPa55w0rdpP-pdaw`

- **Tenant ID**: Enter the `tenant` ID; in the example from the previous chapter, this was `c5df827f-a940-4d7c-b313-426cb3c6b1fe`

Once the form has been filled in, click **Save**. Once saved, you will notice that the **Client Secret** value is marked as **Encrypted**:

Figure 16.7 – Adding our Service Principle to Ansible AWX

When you save sensitive information in Ansible AWX, it is encrypted, and you only have the option to **Replace** or **Revert** it. At no point can you view this information again.

Next, we need to create a credential that contains the private portion of the SSH key we generated in *Chapter 15, Using Ansible with GitHub Actions and Azure DevOps*. To do this click on **Add** again, but this time, enter the following:

- **Name**: AzureVM
- **Description**: Private SSH Key for Azure VMs
- **Organization**: Default
- **Credential Type**: Select **Machine**

In the additional information boxes, enter the following information:

- **Username**: azureadmin
- **Password**: Leave blank
- **SSH Private Key**: Copy and paste the contents of the private key or upload the private key file
- **Remaining options**: Leave blank

Once filled in, click **Save**. Once back on the **Credentials** screen, click **Add** once more and enter the following:

- **Name**: Ansible Galaxy
- **Description**: Ansible Galaxy creds for Default org
- **Organization**: Default
- **Credential Type**: Select **Ansible Galaxy/Automation Hub API Token**

Then, enter this information:

- **Galaxy Server URL**: https://galaxy.ansible.com
- **Remaining options**: Leave blank

Again, click **Save**. Now, it's time to add our final set of credentials:

- **Name**: WordPress Vault
- **Description**: Vault Password for WordPress secrets
- **Organization**: Default
- **Credential Type**: Select **Vault**

In the **Type Details** section, enter the following:

- **Vault Password**: I have added the passwords (which I will tell you later in this chapter) as pre-encrypted Ansible Vault variables in `group_vars/common.yml` in the `Chapter16` playbook. Because of that, you must enter a password of `wibble` here – if you don't enter this, the example playbook will fail.

- **Vault Identifier**: Leave blank.

That was our final credentials. So, let's move on to the next configuration step.

### Adding an inventory

Now that we have all our credentials in place, we need to recreate the content of the `production` inventory file within Ansible AWX. As a reminder, the inventory file we have been using looks like this (minus the comments):

```
[local]
localhost ansible_connection=local

[vmgroup]

[azure_vms:children]

vmgroup
[azure_vms:vars]
ansible_ssh_user=adminuser
ansible_ssh_private_key_file=~/.ssh/id_rsa
host_key_checking=False
```

To add the inventory, click on **Inventories**, which is again in the left-hand menu. The **Add** button now brings up a drop-down list; we want to select **Add inventory** from that list.

In the form that opens, enter the following:

- **Name**: `Azure Inventory`
- **Description**: `Azure Inventory`
- **Organization**: `Default`
- **Instance Groups**: We will add these in a moment
- **Labels**: Leave blank

- **Variables**: Enter the values listed here:

```
ansible_ssh_user: "adminuser"
ansible_ssh_private_key_file: "~/.ssh/id_rsa"
host_key_checking: false
```

Once entered, click **Save**; this will create the inventory. Now, we can add the two groups we need. To do this, click **Groups**, which can be found in the row on the buttons above the details of the inventory:

Figure 16.8 – Adding the inventory to Ansible AWX

Click **Add** and enter the following details:

- **Name**: vmgroup
- **Description**: vmgroup
- **Variables**: Leave blank

Then, click **Save**, repeat the process, and add a second group using the following details:

- **Name**: azure_vms
- **Description**: azure_vms
- **Variables**: Leave blank

Again, click **Save**; you should now have two groups listed.

Now that we have our project, inventory, and some credentials for accessing our Azure environment, we need to add the templates to launch and configure the cluster and terminate it.

## Adding the templates

Let's look at adding the templates.

> **Information**
>
> We will pass a runtime variable to our playbook, which will contain the public part of the SSH key – we added the private portion as a credential earlier in this chapter – and will be called `ssh_key_public`. Please ensure you have the public key when filling out these details.

Click **Templates** in the left-hand menu and, in the drop-down menu of the **Add** button, select **Job Template**. This is the most extensive form we have encountered; however, parts will be populated automatically when we fill in the details. Let's make a start:

- **Name**: `Launch WordPress`
- **Description**: `Launch WordPress in Azure`
- **Job Type**: Select **Run**
- **Inventory**: Select **Azure Inventory**
- **Project**: Select **Azure WordPress**
- **Execution Environment**: Select **AWX EE (latest)**
- **Playbook**: Choose **Chapter16/site.yml** from the drop-down list
- **Credentials**: Select the following:

  - **Machine: AzureVM**
  - **Microsoft Azure Resource Manager: Azure**
  - **Vault: WordPress Vault**

- **Variables**: You should enter the `ssh_key_public` variable here; a truncated version of what to enter is shown here:

  ```
  ---
  ssh_key_public: "ssh-rsa AAAAB3NzaC1yc2EAAAADAQABAAABgQDCGosD-
  5doqnJgOLpkztaDvIZFaCKoChm9yyU6FPaci9fZR60SCXbOu1zeMmyJouFH7xVB-
  v7xw5HBk0FDNLXrssR5B7YHiti8= youremail@example.com"
  ```

- **Remaining options**: Leave blank

Click **Save**; you will be taken to the overview of the template:

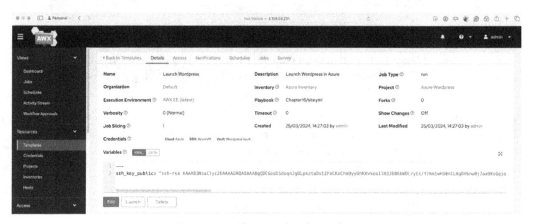

Figure 16.9 – The completed template

Once added, we need to repeat this process with the following details for the Playbook that terminates our deployment:

- **Name**: `Terminate WordPress`
- **Description**: `Terminate WordPress in Azure`
- **Job Type**: Select **Run**
- **Inventory**: Select **Azure Inventory**
- **Project**: Select **Azure WordPress**
- **Execution Environment**: Select **AWX EE (latest)**
- **Playbook**: Choose **Chapter16/destroy.yml** from the drop-down list
- **Credentials**: Select the following:

  - **Microsoft Azure Resource Manager**: **Azure**

- **Remaining options**: Leave blank

Once you've filled in these details, click **Save**.

We have everything we need to run our playbooks, so let's do that.

## Running our playbooks

Back on the **Templates** page, you should see the two templates we have configured listed:

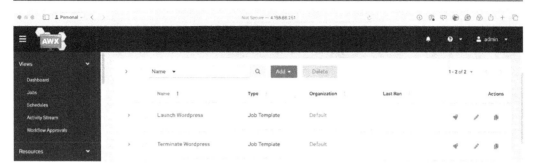

Figure 16.10 – Our two templates

To run the playbook from this page, click on the **Rocket** icon on the **Launch WordPress** template; this will initiate the playbook run and take you to a job page where you will be able to review the status of the playbook job:

Figure 16.11 – Launching WordPress in Azure using Ansible AWX

If everything has worked as planned, after about 5 minutes, you should get confirmation that the playbook has been completed and that your resources have been launched:

Figure 16.12 – Ansible AWX has finished running the playbook

From here, you can re-run the launch playbook again, and it should pick up the newly deployed resources as it did when we re-ran the playbook on our local machine.

Given the number of Azure resources we have launched, before we review the changes to the playbook code to run it in Ansible AWX, we should terminate the WordPress resources. Click on the **Rocket** icon next to the **Terminate WordPress** template to tear down the resources we just launched.

## Terminating the Kubernetes cluster

Before terminating the Azure AKS resources, I recommend clicking around and exploring the Ansible AWX interface. Once you've finished, you can remove the Azure resources and tidy up your local configuration by running the following commands:

```
$ AKSRG=rg-awx-cluster
$ AKSCLUSTER=aks-awx-cluster
$ az aks delete --resource-group $AKSRG --name $AKSCLUSTER
$ az group delete --name $AKSRG
$ kubectl config delete-cluster $AKSCLUSTER
$ kubectl config delete-context $AKSCLUSTER
```

The cluster will take about 5 minutes to remove. To be safe, please don't close any windows until it has finished.

> **Information**
>
> As always, please double-check that your cloud resources have been terminated – you don't want to incur any unexpected costs.

Now that we've terminated all the cost-incurring resources, let's discuss some of the considerations we had to make in our Playbook.

# Playbook considerations

While we touched very lightly on some of the changes that we had to make to our playbook so that it runs on Ansible AWX, let's do a deeper dive now.

## Changes to the existing playbook

As we were running the code locally, to keep the playbook simple, we created a file called `secrets.yml` and loaded the variables from there. Now that we are running Ansible in a shared environment, we should treat our Ansible execution environment as if it were ephemeral, meaning that we cannot rely on this approach.

I used Ansible Vault to encrypt the passwords and ship them within the code to get around this. To do this, I ran the following commands:

```
$ ansible-vault encrypt_string 'SomeP4ssw0rd4MySQL' --name 'db_
password'
$ ansible-vault encrypt_string 'aP455w0rd4W0rDPR355' --name 'wp_
password'
```

When prompted to enter the Vault password, I entered `wibble` as the password, which we then set the Vault password in Ansible AWX when adding credentials. You can see the results of the preceding command in the `group_vars/common.yml` file.

Going back to the playbook code when we ran the playbook from our local machine in *Chapter 9, Moving to the Cloud*, the variable that contains the data for the public SSH key looked like this:

```
vm_config:
  key:
    path: "/home/adminuser/.ssh/authorized_keys"
    data: "{{ lookup('file', '~/.ssh/id_rsa.pub') }}"
```

As you can see, we populate the `vm_config.key.data` variable by reading in the contents of the `~/.ssh/id_rsa.pub` file. However, when we moved our playbook to Ansible AWX, this file no longer exists.

Because of that, we added the `ssh_key_public` variable, which contains the public portion of the private key we uploaded when adding the machine credential when we configured the template. This launches the resources in Azure. This meant the code needed to be updated to the following:

```
vm_config:
  key:
    path: "/home/adminuser/.ssh/authorized_keys"
    data: "{{ ssh_key_public }}"
```

As far as changes go, there's nothing too dramatic and hopefully not unexpected.

### Ansible Galaxy collections

You may not have noticed, but we didn't have to consider the modules that interact with Azure, one of the initial things we covered in *Chapter 9, Moving to the Cloud*.

Ansible AWX does not support these and other collections of modules we need for our playbook to run out of the box, so how did our playbook work without giving an error?

When we first added the project, we configured it to use the GitHub repository that supports this book and contains all the code we have discussed so far. This repository can be found at `https://github.com/packtPublishing/Learn-Ansible-Second-Edition/`.

We only instructed Ansible AWX to use the `site.yml` and `destory.yml` files from the `Chapter16` folder, but in the background, Ansible AWX also used the `requirements.yml` file, which can be found in the `collections` folder in the repository's root.

This file contains the following code:

```
---
collections:
  - name: "azure.azcollection"
    source: "https://galaxy.ansible.com"
  - name: "community.general"
    source: "https://galaxy.ansible.com"
  - name: "community.mysql"
    source: "https://galaxy.ansible.com"
```

As you can see, this is letting Ansible AWX know that it needs to download the `azure.azcollection`, `community.general` and `community.mysql` collections from Ansible Galaxy and, in the background, install their prerequisites.

The only thing we needed to do to get this to work was to create the Ansible Galaxy credential and attach it to our default organization. This means that whenever Ansible AWX comes across a `collections/requirements.yml` file, it will authenticate against Ansible Galaxy using the credentials provided, which in our case were anonymous as we weren't pulling a private collection.

We can also do things such as pin collections to a particular version or add a role:

```
collections:
  - name: "azure.azcollection"
    source: "https://galaxy.ansible.com"
    version: 2.0.0

roles:
  - name: "russmckendrick.learnansible_example"
    source: "https://galaxy.ansible.com"
```

You can also provide different URLs if you are self-hosting an installation of Ansible Galaxy or even provide links to Git repos containing your roles and collections.

This means that Ansible AWX can be as flexible as running Ansible from your local machine.

Before we finish discussing Ansible AWX, let's look at the pros and cons of running it.

## Ansible AWX's advantages and disadvantages

I am sure you will agree from our time with Ansible AWX that it looks like a great tool. However, there are some advantages and disadvantages to running it.

### Open source

Ansible AWX is an open source project, which means it is freely available for anyone to use, modify, and contribute to. This can significantly reduce costs compared to proprietary solutions. However, it has limited Enterprise features.

Ansible AWX offers a good range of features. Still, some advanced enterprise-specific capabilities in Red Hat Ansible Automation Platform, such as advanced reporting, **service-level agreement** (**SLA**) management, and more comprehensive integrations, may be needed.

### Community-driven development

Being open source, Ansible AWX has a strong community of developers and users actively contributing to its development, providing support, and sharing best practices.

However, as an open source project, Ansible AWX relies on community support rather than official commercial support. The community is generally active and helpful, but there are no guaranteed response times or even that someone will be able to help outside of the commercial Red Hat Ansible Automation Platform offering.

### Frequent updates and improvements

Ansible AWX follows a more frequent release cycle than Red Hat Automation Platform. This means that you can gain access to new features, bug fixes, and improvements more quickly.

Ansible AWX's frequent release cycle means you may need to update more often to access the latest features and bug fixes. Upgrading Ansible AWX can require more effort to ensure compatibility and stability, especially in production environments.

Updates and Ansible AWX have always been challenging; they have always been more of a migration than an in-place update.

Using our quick deployment of Ansible AWX as an example, we would need a way to upgrade it. We would have to deploy an external database server outside of our Kubernetes cluster for a more production-like environment – this would contain and persist all our data and configuration.

To *update* Ansible AWX, we would need to tear down all of the resources in the cluster (minus the database), update the AWX Operator, and then redeploy Ansible AWX running the latest version – this would then connect to our external database and run all of the necessary database migration scripts to update our schema and data to make it compatible with the new version.

### Solid foundation

Ansible AWX provides robust features for managing and executing Ansible playbooks, making it a solid choice for organizations starting their automation journey or having more straightforward automation requirements.

### Flexibility and customization

While Ansible AWX integrates with various tools and systems, it may have a different level of out-of-the-box integrations and certified content than Red Hat Automation Platform, which is designed to work seamlessly with other Red Hat products and has a broader ecosystem of supported integrations.

Ansible AWX may also have limitations when managing large-scale deployments or complex enterprise environments. Additional setup, configuration, and resources may be required to handle high-volume automation tasks effectively.

## Summary

This chapter explored Ansible AWX and touched upon Red Hat Automation Platform, two powerful graphical interfaces for managing and streamlining Ansible deployments.

We learned about their differences, the benefits they offer, and how to install and configure Ansible AWX on a Kubernetes cluster in Microsoft Azure. We successfully ran our playbook to launch and terminate WordPress running in Azure using Ansible AWX by setting up a project, credentials, inventory, and templates.

Throughout the process, we discovered the necessary playbook considerations and modifications, such as using Ansible Vault for sensitive information, handling SSH keys, and leveraging Ansible Galaxy collections.

While Ansible AWX offers numerous advantages, including its open source nature, community-driven development, and solid foundation, it is essential to be aware of its potential limitations in enterprise environments and the challenges associated with updating the platform.

The only thing we didn't discuss was the costs of running the commercially supported enterprise-grade Red Hat Automation Platform. Red Hat does not publicly publish them on its website. You must contact one of its partners or Red Hat directly for details.

In our next and final chapter, we will look at some of the ways you can integrate Ansible into your daily workflows, debug your playbooks as they run, and some real-world examples of how I have used Ansible.

## Further reading

To learn more about the topics that were covered in this chapter, take a look at the following resources:

- **Ansible AWX Project**: `https://github.com/ansible/awx`
- **Ansible AWX Operator**: `https://github.com/ansible/awx-operator`
- **Ansible AWX documentation**: `https://ansible.readthedocs.io/projects/awx/en/latest/`
- **Red Hat Ansible Automation Platform**: `https://www.redhat.com/en/technologies/management/ansible`

# 17

# Next Steps with Ansible

In this, our final chapter, we will discuss how you can integrate Ansible into your day-to-day workflows. We will cover continuous integration tools, monitoring tools, and troubleshooting.

We will discuss the following topics:

- Integrating with third-party services
- How you can use Ansible to troubleshoot problems when they occur
- Some real-world examples

Let's dive straight in and look at how we can hook our playbooks into third-party services.

## Technical requirements

This chapter will differ from previous ones. While code examples are given in the chapter and the GitHub repository, they will not be complete working examples. Instead, we will discuss integrating them into your projects so they are more of the art of the possible rather than fully formed examples.

## Integrating with third-party services

Although you may be running the playbooks yourself, it's a good idea to keep a log of your playbook run or update other team members or departments with the results. Ansible has several modules that allow you to work with third-party services to provide real-time notifications.

Let's start by looking at Slack.

### Slack

Slack has rapidly become the preferred option for team-based collaboration services across different IT departments. One key benefit of Slack is its support for third-party applications via its App Directory; Ansible supports Slack Incoming Webhooks via the `community.general.slack` module.

Remember, you can install the `community.general` collection if you don't have it installed by running the following command:

```
$ ansible-galaxy collection install community.general
```

Before we look at the Ansible code, we should quickly discuss how you create a Slack App and enable webhooks.

First, you must make your own Slack app; you can do this by visiting `https://api.slack.com/apps/new`. Once there, click the **Create an App** button and select the **From Scratch** option. From here, you need to fill in the **App Name** and **Pick a workspace to develop your app in**, which for the majority of us will be your primary workspace, as you can see from the following screenshot:

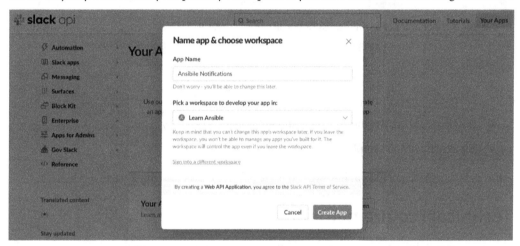

Figure 17.1 – Creating the Slack app

Once the Slack App has been created, you will be taken to your new application settings page. In the left-hand menu, you should see an option for **Incoming Webhooks**. Go to this page and toggle the **Activate Incoming Webhooks** switch to **On**. This will extend the options and give you the option to **Add New Webhook to Workspace**.

From here, you will need to select where you would like your Slack App to post; as you can see from the following screenshot, I selected the **#general** channel:

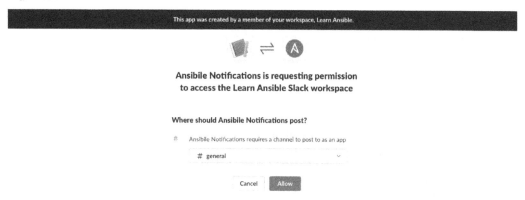

Figure 17.2 – Choosing where to post

Once selected, you will be taken back to the **Incoming Webhooks** page for your application; here, you will be given a Webhook URL, which should look something like the following, and you will need to make a note of this and keep it safe (the following one has been revoked):

```
https://hooks.slack.com/services/TBCRVDMGA/B06RCMPD6R4/
YBTo7ZXZHrRg57fvJXr1sg43
```

Now that we have everything we need to interact with Slack, we can examine the code. As mentioned at the start of the chapter, I will only go into some of the code, as much of it will already be familiar.

There is just a single variable we need to add, and it is the token used to identify and authenticate against the webhook we created: the token is everything after `https://hooks.slack.com/services/` in the webhook URL from Slack, so in my case, the variable, which I put in `group_vars/common.yml`, looks like this:

```
slack:
  token: "TBCRVDMGA/B06RCMPD6R4/YBTo7ZXZHrRg57fvJXr1sg43"
```

As this token should be treated as a secret, I recommend also using Ansible Vault to encrypt the value, so to do this, you can run the following:

```
$ ansible-vault encrypt_string 'TBCRVDMGA/B06RCMPD6R4/
YBTo7ZXZHrRg57fvJXr1sg43' --name 'token'
```

The token in the repo is encrypted using Ansible Vault, and as it has been revoked, you will need to update it with your own.

By jumping straight into `roles/slack/tasks/main.yml`, you can see that the playbook launches a resource group, virtual network, and subnet in Azure.

There are no changes to the first tasks that launch the Azure resources:

```
- name: "Create the resource group"
  azure.azcollection.azure_rm_resourcegroup:
    name: "{{ resource_group_name }}"
    location: "{{ location }}"
    tags: "{{ common_tags }}"
  register: "resource_group_output"
```

Additionally, the debug task we used in previous chapters is still there; immediately after the debug task, we have the task (well, sort of) which sends the notification to Slack:

```
- name: "Notify the team on Slack about the resource group status"
  include_tasks: slack_notify_generic.yml
```

As you can see, it triggers another task in the `slack_notify_generic.yml` file, and we pass the registered output's content as a set of variables, most of the them are self-explanatory:

```
vars:
  resource_changed: "{{ resource_group_output.changed }}"
  resource_type: "Resource Group"
  resource_name: "{{ resource_group_output.state.name }}"
  resource_location: "{{ resource_group_output.state.location }}"
```

The last two are a little different; this one takes the full resource ID and prefixes it with `https://portal.azure.com/#resource`, as the resource ID is the URL for the resource in Azure; this, together with the URL prefix, will give us a clickable link that will take the user directly to the resource when they follow it:

```
    azure_portal_link: "https://portal.azure.com/#resource{{ resource_
group_output.state.id }}"
```

The final variable generates a comma-separated list of tags and values using a Jinja2 template function:

```
    resource_tags: >
      {% for key, value in resource_group_output.state.tags.items() %}
      *{{ key }}:* {{ value }}{% if not loop.last %}, {% endif %}
      {% endfor %}
```

You might also have noticed that the `{{ key }}` variable has a * on either side; this is not part of the template function; this is the markdown syntax for **bold**, and it will style the contents as such.

Before we look at what is in `roles/slack/tasks/ slack_notify_generic.yml`, let's quickly discuss why we are taking this approach.

As we mentioned several times in the title, one of the main goals of automating our deployments is to streamline everything as much as possible. In this case, the task we are calling will be the standard throughout the playbook, and the only changes we need to make are the content.

So rather than repeating the `community.general.slack` task several times in our playbook, we can define it once and then call it multiple times. This means if we need to change something in the `community.general.slack` task, we only have to update it in one place.

The task itself has a little bit of logic added, so let's review that now:

```
- name: "Notify the team on Slack about resource changes"
  community.general.slack:
    token: "{{ slack.token }}"
    parse: "none"
```

As you can see from the preceding code, we are passing our webhook `token` and setting the `parse` option to none. This means that `community.general.slack` will not touch any content we post to the webhook to strip out formatting, etc.

Rather than sending a simple message, we use the `attachments` type. This will nicely format our message into blocks, and we can also set a status color based on whether there has been a change to the content or not:

```
    attachments:
      - fallback: "Notification about Azure resource changes"
```

The logic for setting the color is as follows: here, we use the Boolean value of `true` or `false` that is passed by the `resource_changed` variable. If the variable equals `true`, it means that the resource has been changed, so we set the color to the pre-defined `warning` color, which is orange; otherwise, the color is set to `good`, which is green:

```
        color: "{% if resource_changed %}warning{% else %}good{% endif
  %}"
        title: "Ansible: {{ resource_type }}"
```

Next, we have the message content: here, we are using a similar logic as we did for setting the color based on whether there has been a change to the resource or not:

```
        text: "{{ resource_name }} has been {% if resource_changed %}
  created/updated{% else %}checked (no changes){% endif %}."
```

Finally, we have the fields; each of these displays the information we are passing to the task in a block, apart from one:

```
fields:
  - title: "Location"
    value: "{{ resource_location }}"
    short: true
  - title: "Azure Portal"
    value: "<{{ azure_portal_link }}|View in Azure Portal>"
    short: true
  - title: "Tags"
    value: "{{ resource_tags }}"
    short: false
```

The value of the Azure portal link is a little different; Slack uses **mrkdwn**, a markup language similar to Markdown but with some differences, especially regarding formatting links. As you can see, we are setting this to the following:

```
<{{ azure_portal_link }}|View in Azure Portal>
```

This is the mrkdwn syntax for creating a clickable link. It will link to the URL being passed in the `{{ azure_portal_link }}` variable. The text after the pipe | is the visible text that will appear in the Slack message and act as the clickable link.

When Slack renders this message, it will display **View in Azure Portal** as clickable text. When someone clicks on it, Slack will open the URL in the `{{ azure_portal_link }}` variable, directing the user to the Azure Portal.

Now that we know what the playbook looks like, let's run it:

```
$ ansible-playbook -i hosts site.yml --ask-vault-pass
```

This will prompt you to provide a valuable password and then deploy the resources; in this case, we don't need to know the output of running the playbook and should, instead, turn our attention to Slack itself:

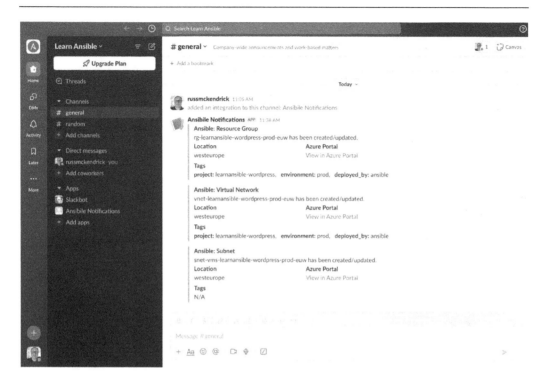

Figure 17.3 – First run of the playbook

As you can see from the preceding output, three resources have been added, so they are referred to as created/updated. The orange bar is on the left-hand side of the message.

Let's now rerun the playbook using the following:

```
$ ansible-playbook -i hosts site.yml --ask-vault-pass
```

You will see that the message now looks like this:

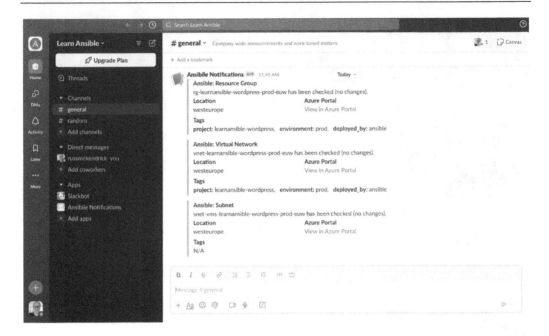

Figure 17.4 – Running the playbook a second time

This time, there have been no changes, which the message reflects. The status is also showing green, so we can quickly see that there have been no changes.

The only thing I would add is that if you look at the code in the repo, you will notice that for the subnet, we are having to make some allowances:

- `resource_location`: subnets don't have a location, so we are using the one from the virtual network the subnet is being created in

- `azure_portal_link`: while an ID for the subnet is being returned, it doesn't precisely match the logic we use to open the resource directly in the Azure portal, so we link to the virtual network where the subnet is configured

- `resource_tags`: you can't add tags to a subnet, so we set the value to N/A

As you can see from the screens, this is useful for notifying others that your playbook is being run. It also gives you quick access to the resources being created/updated or checked and an audit trail of changes being made to your resources.

While the code we discussed only applies to Slack and the resources deployed in Microsoft Azure, the concept should apply to any integration supported by Ansible.

## Other integrations

Dozens of other integrations, both community- and vendor-supported, are available on Ansible Galaxy. If you can't find one for your use case and your target service has an API, you could quite quickly build an integration using the `ansible.builtin.uri` module, which is designed to interact with web APIs and services.

What follows are some example use cases for other integration modules.

### Say

Most modern computers come with some level of voice synthesis built in; by using this module, you can have Ansible verbally inform you of the status of your playbook run:

```
- name: "Speak an update"
  community.general.say:
    msg: "Hello from Ansible running on {{ inventory_hostname }}"
    voice: "Zarvox"
  delegate_to: localhost
```

While this is fun, it isn't very useful and could quickly become annoying, so let's move on.

### Syslog

Suppose you ship the log files from your target hosts. In that case, you may want to send the results of the playbook run to your target host machine syslog so that it is shipped to your central logging service for use in external services such as an **SIEM**, which stands for **security information and event management**, product:

```
- name: "Send a message to the hosts syslog"
  community.general.syslogger:
    msg: "The task has completed and all is well"
    priority: "info"
    facility: "daemon"
    log_pid: "true"
```

This is a great way to register that something has happened on the target host in a way that logs it along with everything else that is happening on the target operating system.

### ServiceNow

ServiceNow is enterprise-grade IT service management software from ServiceNow, Inc.

By using the `servicenow.servicenow.snow_record` module, your playbook can open incidents within your ServiceNow installation:

```
- name: "Create an incident in SNOW"
  servicenow.servicenow.snow_record:
    username: "{{ snow.username}}"
    password: "{{ snow.passord}}"
    instance: "{{ snow.instance }}"
    state: "present"
    data:
      short_description: "Ansible playbook run on {{ inventory_
hostname }}"
      severity: 3
      priority: 2
  register: snow_incident_result
```

Once open, you can then add notes to them using something like the following:

```
- name: "Update the SNOW incident with work notes"
  servicenow.servicenow.snow_record:
    username: "{{ snow.username}}"
    password: "{{ snow.passord}}"
    instance: "{{ snow.instance }}"
    state: present
    number: "{{snow_incident_result['record']['number']}}"
    data:
      work_notes : "{{ resource_name }} has been {% if resource_
changed %}created/updated{% else %}checked (no changes){% endif %}."
```

At the end of the playbook run, you can close the incident, which will permanently record whatever information you ship from your playbook in your ITSM tool.

### Microsoft Teams

While we covered Slack as the primary example in this chapter, Ansible also supports several Microsoft 365 products, including Microsoft Teams, via the `community.general.office_365_connector_card` module. Microsoft 365 Connector cards are very powerful, and their configuration and, by extension, the Ansible module can get quite complicated; so rather than cover them here, I would recommend the following links as a starting point:

- `https://docs.ansible.com/ansible/latest/collections/community/general/office_365_connector_card_module.html`

- `https://learn.microsoft.com/en-us/microsoftteams/platform/task-modules-and-cards/what-are-cards`

- `https://adaptivecards.io/`

As you can see from the preceding links, connector cards can be as simple or complicated as you want. However, configuring them is probably worth a chapter all by itself, so let's move on.

## Summary of third-party services

One of the key takeaways I hope you get from this book is that automation is great; it is not only a real time saver, but using tools such as the ones we covered in the previous chapter, *Chapter 16, Introducing Ansible AWX and Red Hat Ansible Automation Platform*, can enable people who are not sys-admins or developers to execute their playbooks from a friendly web interface. We will look at this further in the final section of the chapter, where I will cover some real-world examples of how Ansible has been implemented in organizations I have worked with.

The modules we have covered in this section allow you to take your automation to the next level by not only allowing you to record the results but also automatically doing some housekeeping during your playbook run and having it notify your users.

For example, you need to deploy a new configuration to your server. Your service desk has made a change request for you to take action on the work within your ServiceNow installation.

Your playbook could be written so that before the change is actioned, it uses the fetch module to copy the configuration file to your Ansible Controller. The playbook could then use the `servicenow.servicenow.snow_record` module to attach a copy of the existing configuration file to the change request, proceed to make the changes, and then automatically update the change request with the results.

Before we look at some real-world examples, let's take a look at how you can debug your playbooks as they are running.

## The Ansible playbook debugger

Ansible has a debugger built in. Let's look at how you can build this into your playbook by creating a simple playbook with an error. As we have just mentioned, we are going to write a playbook that uses the `community.general.say` module. The playbook itself looks like this:

```
- name: "A simple playbook with a mistake"
  hosts: "localhost"

  debugger: "on_failed"

  vars:
    message: "The task has completed and all is well"
    voice: "Daniel"

  tasks:
```

```
    - name: "Say a message on your Ansible host"
      community.general.say:
        msg: "{{ massage }}"
        voice: "{{ voice }}"
```

There are two things to point out: the first is the mistake. As you can see, we are defining a variable named `message`, but when I came to use it in the task, I made a typo and entered `massage` instead. Luckily, as I developed the playbook, I instructed Ansible to use the interactive debugger whenever a task fails by setting the `debugger` option to `on_failed`.

## Debugging the task

Let's run the playbook and see what happens:

```
$ ansible-playbook playbook.yml
```

The first problem is that we are not passing a host inventory file, so there will be warnings that only the localhost is available; this is fine, as we want to run the `Say` module only on our Ansible Controller anyway:

```
[WARNING]: No inventory was parsed, only implicit localhost is
available
[WARNING]: provided hosts list is empty, only localhost is available.
Note that the implicit localhost does not match 'all'
```

Next, Ansible runs the play itself; this should result in a fatal error:

```
PLAY [A simple playbook with a mistake] ********************
TASK [Gathering Facts] ************************************
ok: [localhost]
TASK [Say a message on your Ansible host] *****************
fatal: [localhost]: FAILED! => {"msg": "The task includes an option
with an undefined variable. The error was: 'massage' is undefined.
'massage' is undefined\n\nThe error appears to be in '/Users/russ.
mckendrick/Code/Learn-Ansible-Second-Edition/Chapter17/debugger/
playbook.yml': line 12, column 7, but may\nbe elsewhere in the file
depending on the exact syntax problem.\n\nThe offending line appears
to be:\n\n  tasks:\n    - name: \"Say a message on your Ansible
host\"\n          ^ here\n"}
```

Typically, the playbook run will stop, and you will be returned to your shell; however, because we have instructed Ansible to drop into the interactive debugger, we now see the following prompt:

```
[localhost] TASK: Say a message on your Ansible host (debug) >
```

From here, we can start to investigate the problem a little more; for example, we can review the error by typing the following command:

```
p result._result
```

In Ansible, when using the debug module, the p command is used to prettify the output of a variable or expression. It stands for **pretty** or **pretty-print**. When you use p result._result in an Ansible debug task, it will display the value of result._result in a more readable and formatted way. The p command uses the pprint (**pretty-print**) function from the Python standard library to format the output.

Once you hit the *Enter* key, the results of the failed task will be returned:

```
{'_ansible_no_log': False,
 'failed': True,
 'msg': 'The task includes an option with an undefined variable. The
error was: \'massage\' is undefined. \'massage\' is undefined\n\nThe
error appears to be in \'/Users/russ.mckendrick/Code/Learn-Ansible-
Second-Edition/Chapter17/debugger/playbook.yml\': line 12, column
7, but may\nbe elsewhere in the file depending on the exact syntax
problem.\n\nThe offending line appears to be:\n\n  tasks:\n    - name:
"Say a message on your Ansible host"\n        ^ here\n'}
```

Let's take a closer look at the variables used in the task by typing the following:

```
p task.args
```

This will return the two arguments we are using in the task:

```
{'msg': '{{ massage }}', 'voice': '{{ voice }}'}
```

Now, let's look at the variables that are available to the task using the following:

```
p task_vars
```

You may have noted that we instructed Ansible to execute the setup module as part of the playbook run, so the list of variables available to the task is very long:

```
        'inventory_hostname': 'localhost',
        'inventory_hostname_short': 'localhost',
        'message': 'The task has completed and all is well',
        'module_setup': True,
        'omit': '__omit_place_
holder__7da4853be448a08d857e98fbabe7afe1b7c97d00',
        'play_hosts': ['localhost'],
        'playbook_dir': '/Users/russ.mckendrick/Code/Learn-Ansible-
Second-Edition/Chapter17/debugger',
        'voice': 'Daniel'},
```

As you can see, there is much information about the environment in which our playbook is being executed. In the list of variables, you will notice that all the information gathered by the setup modules starts with `ansible_`, and our two variables are listed at the bottom.

We can find out more about these two variables by running the following commands:

```
p task_vars['message']
p task_vars['voice']
```

This will display the contents of the variable:

```
[localhost] TASK: Say a message on your Ansible host (debug) > p task_
vars['message']
'The task has completed and all is well'
[localhost] TASK: Say a message on your Ansible host (debug) > p task_
vars['voice']
'Daniel'
```

We know we are passing a misspelled variable to the `msg` argument, so we will make some changes on the fly and continue the playbook run. To do this, we are going to run the following command:

```
task.args['msg'] = '{{ message }}'
```

This will update the argument to use the correct variable; we can now rerun the task by issuing the following command:

```
redo
```

This will immediately rerun the task with the correct argument and, with any luck, you should hear, *"The task has completed, and all is well."*

```
changed: [localhost]
PLAY RECAP ***************************************************
localhost: ok=1   changed=1   unreachable=0   failed=0.
skipped=0   rescued=0.  ignored=0
```

As you can see from the preceding output, because we only have a single task, the playbook is completed. If we had more, it would carry on from where it left off. You can now update your playbook with the correct spelling and proceed with the rest of your day. Additionally, if we wanted to, we could have typed either `continue` or `quit` to proceed or stop, respectively.

## Summary of the Ansible debugger

The Ansible debugger is a handy option to enable when you are working on creating complex playbooks; for example, imagine that you have a playbook that takes about 20 minutes to run, but it throws an error somewhere toward the end, say, 18 minutes after you first run the playbook.

Having Ansible drop into the interactive debugger shell not only means you can see precisely what is and isn't defined, but it also means you don't have to blindly make changes to your playbook and then wait another 18 minutes to see whether those changes resolved the fatal error.

# Some real-world examples

Before we finish the chapter and the book, I will give a few examples of how I have used and interacted with Ansible over the last few years.

## Automating a complex deployment

In this example, an application was distributed across several dozen servers in a public cloud. Each application component was installed on at least three different hosts and required updates in a specific order.

The application developers collaborated with the operations team to streamline the deployment process to create an Ansible Playbook. The playbook automated the following steps for each component of the application:

1.  Put the application into maintenance mode by connecting to the targeted hosts and executing a specific command.

2.  Create snapshots of all the costs involved in the deployment, ensuring a rollback point if needed.

3.  Initiate the deployment process by pulling the latest code from the designated GitHub repository and executing a series of commands to update the application.

4.  Verify the deployment's success by connecting to the application's API and running a set of health checks on each targeted host.

5.  If the deployment and health checks pass successfully, take the application out of maintenance mode and proceed to the next component. However, if any tests fail, halt the deployment immediately and execute commands to revert the hosts to the previously taken snapshots, ensuring a safe rollback.

Prior to implementing Ansible automation, the manual execution of these deployment steps took several hours, as the application and operations teams had to co-ordinate and follow the process meticulously. This manual approach made deployments challenging and prone to human errors.

By automating the deployment tasks using Ansible, the teams could focus on handling the exceptions that arose due to genuine issues rather than mistakes caused by manual execution. Before the automation was put in place, errors were common during almost every release, with many hosts and complex manual steps involved.

The introduction of Ansible automation significantly improved the deployment process, reducing the time required and minimizing the risk of human errors. The playbook ensured consistency, reliability, and repeatability across multiple deployments, enabling the teams to deploy the application components more frequently and with greater confidence.

This example demonstrates how Ansible can tackle complex deployment scenarios, streamline processes, and enhance collaboration between development and operations teams in a public cloud environment.

## Combining Ansible and other tools

In this real-world scenario, we collaborated with a team that had invested significant effort in developing their infrastructure automation using Terraform. Their Terraform code successfully deployed the infrastructure and performed basic host bootstrapping using a simple `cloud-init` script.

However, as the application requirements grew more complex, it became evident that additional automation was needed to effectively manage the application on the provisioned hosts. Instead of replacing the existing Terraform code, we introduced Ansible to complement the infrastructure automation.

To integrate Ansible with the existing Terraform workflow, we utilized the `community.general.terraform` module. This module allowed us to execute the Terraform deployment directly from within an Ansible playbook.

By leveraging this integration, we took the output generated by the Terraform deployment and passed the relevant information back to Ansible. This enabled Ansible to gather detailed information about the provisioned hosts and perform the necessary application bootstrapping tasks.

The combination of Terraform and Ansible proved to be a powerful solution:

- Terraform handled the infrastructure provisioning, ensuring the required resources were created and configured correctly in the target environment.

- Ansible took over the application management, utilizing the host information provided by Terraform to configure and deploy the application components seamlessly.

This approach allowed the team to maintain their existing Terraform codebase while extending the automation capabilities with Ansible. The integration between the two tools provided a seamless workflow, enabling the team to manage both the infrastructure and the application more effectively without having to throw away the code that they already had.

The team achieved a more comprehensive and efficient automation solution by choosing the right tools for specific tasks and leveraging their strengths. Terraforms infrastructure-as-code capabilities, combined with Ansible's application management and orchestration features, resulted in a robust and flexible automation pipeline.

## Deploying Ansible AWX

As discussed in *Chapter 16, Introducing Ansible AWX and Red Hat Ansible Automation Platform*, Ansible AWX is a powerful tool that offers a wide range of features beyond the basics. In addition to the core functionalities, Ansible AWX provides capabilities such as surveys, integration with identity services such as Microsoft Entra, and **role-based access controls** (**RBACs**) that enable granular access management for projects and templates.

Surveys in Ansible AWX allow you to create interactive forms that gather input from users before running a playbook. This feature is particularly useful when you need to collect specific information or parameters from end-users without exposing them to the underlying playbook complexities.

Integration with identity services, such as Microsoft Entra, enables seamless authentication and authorization for Ansible AWX users. This integration allows you to leverage existing user accounts and access controls, simplifying user management and ensuring secure access to Ansible AWX resources.

RBAC in Ansible AWX provides a flexible and granular way to manage user permissions. With RBAC, you can define roles and associate them with specific projects, templates, and other resources. This allows you to control who can access and execute specific playbooks, ensuring that users have the appropriate level of access based on their responsibilities and expertise.

In the following examples, we'll explore how Ansible AWX has been utilized in various organizations that I have worked with to streamline processes, automate tasks, and empower teams to perform their duties effectively while maintaining security and governance.

## Provisioning virtual machines

In this scenario, the IT team needed to provide a self-service portal for developers to provision **virtual machines** (**VMs**) across different environments, such as development, staging, and production. Each environment had specific requirements and configurations.

To streamline the process, Ansible AWX was deployed, and a survey was created to capture the necessary information from the developers. The survey included fields for specifying the desired operating system, VM size, environment, and other relevant parameters.

Upon submitting the survey, Ansible AWX triggered a playbook that automated the provisioning process. Based on the survey responses, the playbook dynamically generated the appropriate VM configurations and provisioned the VMs in the specified environment.

Additionally, the playbook integrated with the organization's ticketing system, automatically creating a ticket with the VM details and linking it to the change management process for tracking and auditing purposes.

By leveraging Ansible AWX and surveys, the IT team empowered developers to provision VMs on-demand while maintaining control and governance over the process.

## Managing application deployments

In another use case, a software development team needed to deploy their application across multiple environments, including development, QA, and production. Each environment had its own set of configurations and dependencies.

To simplify the deployment process, Ansible AWX was utilized. A survey was created to capture the necessary deployment parameters, such as the application version, target environment, and any specific configuration options.

The survey responses were then passed as variables to an Ansible playbook that was responsible for executing the deployment. The playbook handled the entire deployment process, including the following:

- Retrieving the specified application version from the artifact repository
- Configuring the target environment based on the provided parameters
- Deploying the application components and dependencies
- Running post-deployment tests and health checks
- Updating the deployment status in the organization's project management tool

By using Ansible AWX and surveys, the development team could initiate deployments through a user-friendly interface, ensuring consistency and reducing the risk of manual errors. The playbook automated the complex deployment steps, saving time and effort for the team who needed the deployment while freeing up the time of the team who would have done the deployment.

## Updating DNS records

In this example, the organization managed multiple **DNS** (or, to give it its full name, **domain name system**) zones across different providers, and they needed to allow front-line support teams to update DNS records without granting them direct access to the providers' management consoles.

To achieve this, Ansible AWX was used. A survey was created to capture the necessary information for updating DNS records. The survey included fields specifying the domain name, record type (e.g., A, CNAME, MX), record value, and **time to live** (TTL).

Upon submitting the survey, Ansible AWX triggered a playbook that automated the DNS record update process. The playbook performed the following steps:

1. Validated the provided survey inputs to ensure data integrity and prevent invalid entries
2. Determined the appropriate DNS provider based on the domain name specified in the survey
3. Connected to the DNS provider's API using the necessary credentials securely stored in Ansible Vault
4. Retrieved the existing DNS records for the specified domain and record type

5. Updated the DNS record with the new value and TTL provided in the survey

6. Saved the updated DNS record using the provider's API

7. Logged the change in the organization's change management system, such as ServiceNow, for tracking and auditing purposes

By using Ansible AWX, the front-line support teams could easily update DNS records without requiring direct access to the DNS providers' management consoles. The playbook automated the complex steps involved in updating DNS records across multiple providers, ensuring consistency and reducing the risk of errors.

Additionally, the integration with the change management system provided a centralized record of all DNS changes, enabling easy tracking, auditing, and compliance with the organization's change control processes.

These examples demonstrate how Ansible AWX can be leveraged to run tasks and simplify processes for end-users across different domains, such as infrastructure provisioning and application deployment. By combining Ansible AWX with surveys and integrating with existing tools and processes, organizations can enable self-service capabilities while maintaining control and governance over critical operations.

# Summary

We have reached the end of the chapter and our book. I have been trying to think of a way to summarize Ansible; I believe the summary from the first edition of *Learn Ansible* still stands.

In response to a technical recruiter who reached out to him with a job role that required at least three years of Ansible experience when the tool had only been available for a short time, Ansible creator Michael DeHaan said the following in a now-deleted Tweet:

*"Anyone using Ansible for a few months is as good as anyone using Ansible for three years. It's a simple tool on purpose."*

That perfectly sums up my experience of Ansible and hopefully yours.

Once you know the basics, it is straightforward to move on and start building more complex playbooks quickly. These playbooks can assist with deploying basic code and applications as well as complex cloud and even physical architectures.

Reusing your roles and accessing an extensive collection of community-contributed roles and modules via Ansible Galaxy means you have many examples or quick starting points for your next project. So, you can roll your sleeves up and get stuck in a lot sooner than you would with other tools. Additionally, if Ansible cannot do something, the odds are that there is a tool it can integrate with to provide the missing functionality.

Going back to what we discussed back in *Chapter 1, Installing and Running Ansible*, being able to define your infrastructure and deployment in code in a repeatable and shareable way that encourages others to contribute to your playbooks should be the aim of starting to introduce Ansible into your day-to-day workflows.

Through this book, I hope you have begun to think of day-to-day tasks where Ansible could help you and save you time, and I wish you luck with developing your own playbooks.

# Index

## A

# C

`packtpub.com`

Subscribe to our online digital library for full access to over 7,000 books and videos, as well as industry leading tools to help you plan your personal development and advance your career. For more information, please visit our website.

## Why subscribe?

- Spend less time learning and more time coding with practical eBooks and Videos from over 4,000 industry professionals
- Improve your learning with Skill Plans built especially for you
- Get a free eBook or video every month
- Fully searchable for easy access to vital information
- Copy and paste, print, and bookmark content

Did you know that Packt offers eBook versions of every book published, with PDF and ePub files available? You can upgrade to the eBook version at `packtpub.com` and as a print book customer, you are entitled to a discount on the eBook copy. Get in touch with us at `customercare@packtpub.com` for more details.

At `www.packtpub.com`, you can also read a collection of free technical articles, sign up for a range of free newsletters, and receive exclusive discounts and offers on Packt books and eBooks.

# Other Books You May Enjoy

If you enjoyed this book, you may be interested in these other books by Packt:

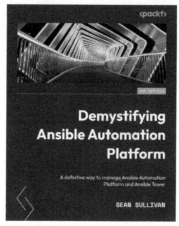

**Demystifying Ansible Automation Platform**

Sean Sullivan

ISBN: 978-1-80324-488-4

- Get the hang of different parts of Ansible Automation Platform and their maintenance
- Back up and restore an installation of Ansible Automation Platform
- Launch and configure basic and advanced workflows and jobs
- Create your own execution environment using CI/CD pipelines
- Interact with Git, Red Hat Authentication Server, and logging services
- Integrate the Automation controller with services catalog
- Use Automation Mesh to scale Automation Controller

**Ansible for Real-Life Automation**

Gineesh Madapparambath

ISBN: 978-1-80323-541-7

- Explore real-life IT automation use cases and employ Ansible for automation
- Develop playbooks with best practices for production environments
- Approach different automation use cases with the most suitable methods
- Use Ansible for infrastructure management and automate VMWare, AWS, and GCP
- Integrate Ansible with Terraform, Jenkins, OpenShift, and Kubernetes
- Manage container platforms such as Kubernetes and OpenShift with Ansible
- Get to know the Red Hat Ansible Automation Platform and its capabilities

## Packt is searching for authors like you

If you're interested in becoming an author for Packt, please visit authors.packtpub.com and apply today. We have worked with thousands of developers and tech professionals, just like you, to help them share their insight with the global tech community. You can make a general application, apply for a specific hot topic that we are recruiting an author for, or submit your own idea.

## Share Your Thoughts

Now you've finished *Learn Ansible*, we'd love to hear your thoughts! Scan the QR code below to go straight to the Amazon review page for this book and share your feedback or leave a review on the site that you purchased it from.

https://packt.link/r/1835088910

Your review is important to us and the tech community and will help us make sure we're delivering excellent quality content.

# Download a free PDF copy of this book

Thanks for purchasing this book!

Do you like to read on the go but are unable to carry your print books everywhere?

Is your eBook purchase not compatible with the device of your choice?

Don't worry, now with every Packt book you get a DRM-free PDF version of that book at no cost.

Read anywhere, any place, on any device. Search, copy, and paste code from your favorite technical books directly into your application.

The perks don't stop there, you can get exclusive access to discounts, newsletters, and great free content in your inbox daily

Follow these simple steps to get the benefits:

1. Scan the QR code or visit the link below

https://packt.link/free-ebook/9781835088913

2. Submit your proof of purchase
3. That's it! We'll send your free PDF and other benefits to your email directly